WILDLIFE
THROUGH THE CAMERA

WILDLIFE THROUGH THE CAMERA

Introduction by David Attenborough

Series producer WILDLIFE-ON-ONE Peter Bale

Having, as it were, gone for knowledge, what one
mustn't do is destroy the wonder, the excitement,
the glamour and joy of the subject.

Desmond Hawkins, speaking in the 'Wildlife Jubilee' programme
celebrating the BBC Natural History Unit's 25th anniversary, 1982

THE BRITISH BROADCASTING CORPORATION

CONTENTS

Published by the British Broadcasting Corporation
35 Marylebone High Street London W1M 4AA

This book was designed and produced by the
Oregon Press Limited, Faraday House,
8-10 Charing Cross Road, London WC2H 0HG

ISBN 0 563 20069 3

First published 1982

© The British Broadcasting Corporation 1982

Design concept: Peter Saag
Art editor: Gail Engert
Design assistant: Graham Keen
Reader: Raymond Kaye

Filmset by SX Composing Ltd,
Rayleigh, England

Printed and bound in Italy by
Amilcare Pizzi s.p.a., Milan

HALF TITLE *Lions at dawn (Ambush at Masai Mara)*
FRONTISPIECE *With its head and thorax out, the*
dragonfly hangs upside down from its abdomen and rests
for up to an hour (The Dragon and the Damsel, p. 171)

INTRODUCTION

Why are so many people interested in animals? Why should twelve million television viewers, in Britain alone, regularly watch 'Wildlife on One' and so put it in the lists of the most popular programmes, alongside quiz shows, domestic serials and light entertainment spectaculars?

Animals, of course, are very beautiful – but not all of them. Few, except the most partisan viewer, could call the blood-lapping vampire bat beautiful, yet a programme about it attracted a vast audience. You might argue that when so many who appear on television are only too aware of the camera and put on special performances for its benefit, it is a relief to watch creatures that behave completely naturally. The American poet Walt Whitman compiled a whole catalogue of reasons why animals, rather than human beings, appealed to him: 'not one', he wrote, 'is demented with the mania of owning things; not one kneels to another'. But then presumably he had never watched a bower bird gathering its treasures and stealing from a rival's collection; and he obviously knew nothing about the social disciplines of a troop of baboons.

When I am asked how I first became interested in animals – sometimes in tones that suggest that this was some kind of aberration from normal and respectable behaviour – I usually stand the question on its head and return it to the questioner. How could anyone, I enquire, have *lost* an interest in animals, since it is clear that all of us, as children, were once fascinated by them? And if you doubt the truth of the statement in the last part of that question, just watch a young child gravely observing a snail as it trundles across a paving stone and prodding it experimentally to see how it withdraws its tentacles; or recall the face of a small boy, incandescent with delight, as he returns from a pond with a jam-jar full of water beetles and gasping sticklebacks.

Perhaps from our earliest years, we recognize that all life is kin, that we and all other animals are ultimately related to one another: a fact that some religions have affirmed, others denied, and that science has only recently demonstrated. So living creatures fascinate us because even the most distantly related of them have problems that we can comprehend and to some extent share – problems of feeding and defence, of mating and ensuring the survival of their offspring. The fact that Saturn consists of poisonous clouds of methane and ammonia interests me and I look at film of it with wonder – for a short time. But for enduring pleasure and continuous interest, I would rather watch a flock of flamingos feeding in the dawn on an African salt lake; and now that astronauts have established that the surface of the moon is covered by deserts of dust that are totally lifeless, that remote and extraordinary place holds less fascination for me than the jungle of South America, filled as it is by monkeys and hummingbirds, toucans and butterflies and myriads of tiny creatures that are still unknown to science.

The conviction that the living world holds an abiding interest for a vast number of people persuaded the BBC, 25 years ago, to set up its Natural History Unit. Since then, the Unit, from its base in Bristol, has produced a long line of programmes both on radio and television. 'Wildlife on One' is one of the more recent. The series was given a wide editorial brief. It was not to concentrate on furry, cuddly creatures even if these do have an obvious and immediate appeal, but to look at all parts of the animal kingdom. So it has filmed not only water voles and dormice but adders and scorpions. It was to feature not only the dramatic and the exotic, but the familiar and domestic. So as well as a programme about lions in Africa, it made another in the English countryside about hedgehogs and, perhaps surprisingly, it was the second that produced the more unusual and revealing sequences.

'Wildlife on One' cast its net widely and its catch was richer than any of us could have hoped. This book has been written by the casters of that net, the producers and directors of the series. It was my continuing and regular pleasure to speak the commentaries for their programmes and it is my privilege now to introduce their accounts not only of what they saw, but of how, with an astounding skill, ingenuity and patience they seldom mention, but which can be detected between their lines, they managed to translate their visions of the wild into television.

David Attenborough

1 Two-hundred year-old beech tree
2 Scratching post used for sharpening claws
3 Badger pair play at one of the entrances to the sett
4 The tunnel system in a badger sett runs for many

yards underground
5 Cubs will spend most of their first weeks in the nest
6 Excavated soil is scaped to the surface and
deposited in heaps around the tunnel entrances

1. AN EYE IN THE NIGHT

BADGERWATCH *producer Peter Bale*

It is often claimed that television is at its best when covering live outside broadcasts where the cameras are able to focus on the action of the moment and highlight the uncertainty and suspense of unknown incidents to come. Not so with the film camera; even Eric Ashby's masterful presentation of the badger is a film record of past history – in television jargon, 'an action replay' of events that had happened over months or even years. That Eric's keen observations had added another dimension to our appreciation of the wild world in action, goes without saying. But his film was once-removed from the first-hand experience of real-life badger watchers, whose patient persistence allows them the privilege of observing badgers with their own eyes.

Badger watching is a rare and specialized human activity. It is a hobby that verges on the obsessive among the small band of enthusiasts who spend all hours of the night, in all kinds of weather, hoping to catch a glimpse of badgers going about their nocturnal business. On one occasion only have I known the excitement of a wild confrontation with Brock, to give him his old country name. It happened in late summer in a copse on the edge of farmland, just a few miles from Bristol on the borders of Somerset and Avon: one of the areas in Britain where the badger is still numerous and his setts most easily tracked down. The last traces of sunset were barely visible across the River Severn, and only a faint glimmer of light penetrated the criss-cross of twigs and branches

that interrupted my view into the copse where our quarry reigned supreme.

My companion insisted on total stillness, absolute concealment and no conversation, even in muffled whispers. After two hours in cramped inconvenience, cold and crumpled by the thicket that surrounded us, I had convinced myself that this badger watching was a fool's game, and that

Television badgerwatching

badger watchers were, if not 'nuts', then misguided to the point of absurdity. But, as is so often the case with city folk, our senses deadened by the cacophony of machinery and the excessive decibels of the highways, I had failed totally to detect the delicate activity that was taking place at that instant in the soft carpet of grass right at my feet. I gazed into the dark, frozen to the ground, as two blazes of white muzzle twitched from side to side, followed by a dark outline that paused only inches away, then shambled on again into the night.

I learned afterwards that this brief event had occurred in the space of seven minutes; it seemed like seven seconds – such is the effect of a sudden new awareness of raw nature. Later that night, still high on the elation of the experience and warmed by an intake of BBC coffee, I promised myself that the Natural History Unit must find some way of televising the instant reality of such moments.

Television producers are often asked how programmes first come about: 'Are they suggested by you, or by scientists?' The answer is, as often as not, much as I have descirbed my own initiation to the marvel of badger life. The idea is not, as some may imagine, conceived by a remote committee in the towers of the Television Centre. It is usually motivated by the personal urge of the individual who is inquisitive by nature and stubborn enough to go on asking 'Why?' My total lack of comprehension on that first nocturnal adventure had certainly stirred my imagination enough to push the creative urge into action. My own interest was awakened on that night so, I suppose, that is when television badger watching more or less began! I do not really know the answer, but then, who really knows what makes the stuff of television except

that you sense it – rather as the badger senses his territory and decides to build his sett, for no better reason than that it seems right and proper to do so, right there and then!

It is much to the credit of Mick Rhodes, then head of Bristol's Natural History Unit, that in spite of the technical difficulties of mounting a live series of transmissions at night from the middle of the Gloucestershire countryside, he was as intrigued as myself at the possibilities of an entirely new technique in bringing nature to the television screen. It was Mick's forceful persuasion that got the project under way with a commitment to transmit a week-night series of 15-minute live shows in the following year on the BBC's main channel. 'You've got the airtime,' I was told, 'now go away and do it, but for goodness sake get it right and make sure we see some badgers!' I admit to having felt a sudden surge of alarm at the thought of actually having to prove that the idea would work.

In the late summer of 1976 I set about seeking advice from the experts, to find answers to the two key questions: where was the best place to televise wild badgers? What problems would arise in having to place cameras within 6 m (20 ft) of a badger sett in order to have an uninterrupted view of them above ground? Dr Ernest Neal is an authority on Brock. His lifetime of research has given him a unique and personal understanding of badger

ABOVE *Commentary caravan* LEFT *Two blazes of white muzzle twitched from side to side*

behaviour. His book, *Badgers*, gave an unfailing source for reference, and at our first meeting at his home in Somerset he jumped at the chance of joining us as an expert commentator – a task he would eventually share with another badger 'buff', Phil Drabble, when the time came for transmission in the following spring.

Ernest put us in touch with Dr Chris Cheeseman, a scientist with the Ministry of Agriculture who was doing fieldwork in Gloucestershire with the controversial 'badger-gassing' programme. The badger's association with the spread of tuberculosis among both domestic herds and wild animals was then, as now, a burning issue; one that seemed endlessly controversial and would have to receive attention within the programme. That we would be commenting on the dilemma, on the air, would enable those less well informed about the conflict between conservationists and those responsible for the health of our domestic animals, particularly cattle, to realize the vulnerability of the badger's future in the wild.

Chris Cheeseman, the man from the Ministry, was a key contact. His work in an isolated, lush Cotswold valley provided the confirmation we needed to guarantee badger activity. He knew his badger families, and was able to pinpoint the setts and their tunnel entrances, so that we could site our cameras with a commanding view over likely areas of surface activity. The location itself was right off the beaten track and approachable only along a dirt lane, but with sufficient level space at the end on which to park our outside broadcast equipment and the radio link vehicles with which to relay pictures into the transmission network. The fact that we had to arrange for many tons of hardcore to be bulldozed onto the track to make it safe for our heavy outside broadcast vehicles seemed a small price to pay for the uninterrupted and totally ideal view we would have of our badgers. Our badgers, they became; and through the winter months leading up to the birth of new badger cubs, each planning visit was carried out with the kind of care reserved for the most valuable of television stars. Only one person at a time was allowed on the sett area, and we made every effort to keep noise and man's physical presence to the minimum. We knew that to disturb our badgers would mean the end of the project. Thankfully, even the most potentially upsetting intrusion of all, the installation of camera scaffolding towers over the sett, was achieved without any noticeable objection from the unsuspecting residents.

Now, the most challenging problem had to be resolved: if our badgers were to remain unsuspecting (and they had to be if our transmissions were to live up to the claim of being genuinely 'true to life') then we would have to restrict human presence even further, particularly in the week of transmission and the days leading up to the broadcasts. This raised the question of what we were to do about the cameraman.

It is the cameraman's job smoothly to control camera movement; to pan, to tilt, to zoom in or out, as well as focus the lens on the subject in his view. To do this he must be able to converse, at all times, with his technical colleagues and the producer in the control vehicle a quarter of a mile away. To add to the complications, we needed to observe badger comings and goings without a flood of light to disturb their moonlight excursions. As every schoolboy knows, there are no pictures without light, so the idea did not exactly endear this producer to his technical colleagues.

Previous experience had shown that the badger, like all wild mammals, will behave naturally only when human beings are absent, choosing to remain underground at the mere hint of human conversation. On the face of it, such conflicts of interest were overwhelmingly in favour of our badgers not being seen on television! So it would have remained, but for the endeavour and expertise of the BBC's technical team. Their challenge was to create enough invisible light for the cameras to see over an area the size of two tennis courts. Also they had to devise a robot that would allow our two cameras to be operated from the control vehicle at the other side of the valley, out of sight and out of mind of our precious badgers.

There is no doubt that the advent of television's 'eye in the night' was a major technical advance in its ability to observe natural events under conditions roughly equivalent to bright moonlight. Looking back on those strenuous months of preparation it must be said that the 'eye' would not have so much as 'winked' without the electronic skill of the engineers. They had been asked to achieve the impossible, but I make no apologies for making remarks like 'I agree, it is too much to ask, it cannot possibly be done . . .' Such comments are

the lowest of low tricks but as my colleagues later admitted, it stimulated action!

If their task was unusual, the manner in which the project team approached it was, to say the least, unconventional. Their machinations caused eyebrows to be raised among those in authority – what was that man from the BBC doing searching in the junk shops off London's Portobello Road? Well, in the absence of any suitable infra-red light sources, John Noakes of the Technical Investigations department in London, with his Bristol colleague Paul Townsend, had set out to rummage among the bottomless treasure chests of government surplus stores. From the electronic depths of Shepherd's Bush market they found the vital cooling fans for the lights. Purchased at bargain prices, their prizes were bolted alongside the filtered light sources in boxes, more like biscuit tins than the two-kilowatt lamps they purported to be. A redundant monochrome camera that had seen better days was dusted down, taken apart and reassembled with a silicon diode videcon tube just powerful enough to amplify the faint images we would see when the cameras were hoisted up into place among the tree branches. As both cameras would need to be exposed to the weather for several weeks, they were wrapped in plastic bags to keep out the damp

LEFT *and* RIGHT *Infra-red remote controlled cameras*

and the dew – essential precautions if we were to keep our unwritten bargain with the badgers to keep off their patch as much as possible.

Meanwhile, Caroline Weaver and John Borley proceeded with research into badger behaviour and the planning of the technical side of the operation. For John in particular, 'Badgerwatch' was loaded with special problems. Apart from the technical side, as Engineering Manager he had to plan staffing requirements so that we could work the long hours most economically and efficiently. The crew would have to be fed on site, then taken home after each late-night transmission had gone out and brought back again the next day in time to record any badger behaviour that might occur before sunset – just in case we were rained out, or otherwise frustrated by the non-appearance of badgers during the transmission periods after dark. All too soon, May had arrived. The gear was in place and working well. 'Badgerwatch' was set to go!

Whenever television chances its arm with something new, it becomes an enterprise of considerable importance for those involved at the sharp end of the camera. If viewers could have seen us on that first night, they might well have wondered what all the fuss was about: there is something ridiculous

about being in an electronic hothouse, surrounded by monitors, cocooned within an air-conditioned box on wheels in the middle of an empty field, shrouded in darkness just waiting for some badgers to show themselves. It is a kind of madness.

'*Ten seconds to go, first pictures please . . .*' I look up at the screens. Our pictures are, we think, spectacular (but then, we are committed enthusiasts).

'*Presentation are showing the symbol – announcing now . . .*' As the network symbol rolls off the screen the first images of the Cotswold night come on the air. You wonder if the pictures and the atmosphere will be powerful enough to project the excitement you feel to those watching at home, remote from the point of action.

'*Cue Bruce Parker, on Camera One, pan slowly past the beech tree . . . hold it there One . . .*' As Bruce Parker takes his cue to welcome viewers to the Gloucestershire valley, there is that flash of doubt that we will fail, horribly!

'*On Camera Two, pan slowly, keep looking . . .*' I snap out of despair – time later for recriminations! Suddenly there is a grunt from beneath the 200-year-old beech tree. There is a rustle in the bed of wild garlic, and a weird shape looms up from underground.

ABOVE, BELOW and RIGHT
*Night images from the
remote controlled
cameras*

'Zoom in One and stay with it.'

The night images from the remotely controlled cameras are a ghostly blue, unreal yet entirely real.

'Hold that shot One . . . follow if you can.' It is real enough. The first wild badger ever seen live on the air, and within seconds of our first transmission. It trundles up the bank, hidden momentarily in the leaves of garlic. The badger pauses and turns as more grunts are picked up by microphones concealed inside the tunnel entrance.

'Good gracious! Just look at that!' Phil Drabble almost explodes with spontaneous excitement. Worries pass, forgotten. With all eyes on the screens, it is the badgers who take over, totally unconcerned but utterly fascinating.

So the pattern developed night by night as, with recordings and live pictures, we built up a simple diary of family life; of skirmishes between animals and those rare moments when the sow emerged from the sett followed by her pair of playful cubs. Remarkably, these incidents were having an unexpected effect on our viewers.

Next morning in the remote seclusion of our empty field we received messages of appreciation, even enquiries about the well-being of a particular creature. We pondered over the centre-spread picture in a popular daily tabloid; it featured the owl chick, a picture taken off the screen just a few hours before. It showed the chick struggling to climb vertically back to its nest in the beech tree from where it had fallen. It had been the chance pan of a camera that first revealed the two chicks at their nest entrance their four bright eyes peering out at the camera. Maybe this baby had fallen in an attempt to fly from the nest. What is certain is that our coverage of the chick regaining a perch high in the tree canopy had, so to speak, stolen the badgers' thunder on their opening night on television!

Happily, on subsequent nights, the badgers were to have the screens all to themselves. They won praise and applause by just being themselves, at a badger sett in a field, somewhere in the Gloucestershire countryside.

In the early hours of Saturday morning, after four nights of transmission, we shut off the power and began dismantling the cameras, knowing that many badger-watching viewers had stayed up with us to enjoy these engaging animals. I still like to believe that our broadcasts did convey to the viewers something of the motivation I had known – just once – on that summer night in the copse near Bristol.

'What do we do now?' the crew asked. 'We've done the badger, what about having a go at master fox?' Well . . . why not?

AT HOME WITH BADGERS *producer Caroline Weaver*

Down in the forest something stirred . . .

No one knows better than Eric Ashby that the badger is one of the most difficult of British wild animals to photograph in its natural setting. A badger's home, or sett, is underground and as a badger family seldom leaves its subterranean refuge to feed before dusk, it is elusive enough to deter all but the most determined of film makers. It was a challenge that would tax Eric's skill and demand those qualities of creative ingenuity with which only the most dedicated wildlife photographers are endowed.

Eric had taken his first wild badger photograph with a still camera and primitive flash attachment, as long ago as 1938, at night in a wood near his New Forest home – something of an achievement at the time – but it was to be 40 years on, almost to the day, before 'At Home With Badgers', his first full television record of the wild badger, would be ready for screening. During that time he would make many other films about New Forest wildlife so, over the years, he came to know the position of some 200 badger setts and gained considerable knowledge about how best to film Brock's all-too-brief appearances above ground.

With a group of perhaps a dozen setts from which to make his final selection, Eric then considered such vital factors as wind direction and the risk of an overhanging tree branch cutting out too much sunlight from the sett entrance to make filming a possibility. He would need maximum available light to get a tolerable exposure, so he had to choose a sett where the badger community tended to leave as early as possible in the evening to begin their search for food. Only when his very specific needs had been exactly met would this lone cameraman set off into the Hampshire countryside suitably armed with camera and tripod.

Usually a badger sett is located within a gently sloping bank protected by clumps of bushes and undergrowth. Badgers often make their network of

ABOVE *It is said that the name 'badger' is derived from the French word* bécheur LEFT *Eric Ashby.*

16

tunnels by excavating earth from between the roots of well-established trees. It is the scraping action of their strong short legs and long claws that makes them experts at building their extensive tunnel system. Each tunnel is about 23 cm (9 in) in diameter, just enough for the 11 kg (25 lb) adults to have easy access. They widen the tunnels to about 30 cm (1 ft) at the entrances, of which there are several sited at convenient points on the ground surface. It is said that the name 'badger' is derived from the French word *bêcheur* meaning digger, which seems entirely appropriate for an animal of such mining ability.

Generally speaking, badgers live in loose family groups, in several adjacent setts, and will often use outlying ones within a given territory as resting places or even temporary homes. During winter they will spend most of the time safe and warm below ground in nest areas built by broadening sections of the tunnel. Lined with dry grass and leaves these small but compact nests make ideal resting places for the long cold days. The best time for a daylight sighting of badgers above ground is when spring comes and the days lengthen. It was on just such days in late April and May that Eric was able to focus his camera on the stars of his film.

His approach to filming wildlife has always had that touch of astute simplicity: he would always arrive at the sett well before the badgers were likely to emerge and, taking care not to walk over the sett itself, he would place his camera and tripod downwind so that they would not scent his presence with their sensitive nostrils. Their eyesight is poor, so he would watch close by, motionless, for hours if necessary, until he could see a badger nervously sniffing the air at the tunnel entrance. On another day there would be a new sound – the high-pitched chittering of young ones, a sure signal that badger cubs were close behind the mother. Adults are very wary with young cubs and only bring them above ground when they are seven or eight weeks old, usually in the safety of darkness. On this occasion it was late afternoon with warm still air to greet the cautious family. Within minutes Eric was filming the cubs in tumbling play with their soft grey fur fluffed up in mock fights between brothers and sisters – a testing ground for the conflicts of adulthood. Usually there is one litter of two or three cubs in each sett. Here there were at least five cubs, which indicated

that two litters may have come together temporarily, before one of the sows moved her family to another sett – probably in search of more food.

With hawk-eyed tenacity, Eric managed to capture, reel by reel, the detail of the badgers' life above ground; the grooming behaviour with the family huddled in a tight bundle removing parasites and dead skin, then licking, pulling, rolling and rubbing each other with the scent gland that identifies each individual within a community and by which they recognize each other in the dark. A sow suckles her cubs for at least 12 weeks after birth and it is usually 15 weeks before the cubs are independent enough to join parents as they venture off along the well-used and clearly defined tracks that lead from the sett to their hunting grounds nearby.

Badgers stick to their tracks with determination and they mark their passage through the undergrowth with the distinctive 'musk' gland that identifies their progress along the way. They spend many hours each night searching for food. They are omnivorous and depend on their sense of smell to find roots, larvae, carrion and insects which they devour with relish. By and large, they adapt to locally available food and their eating habits seem to be helpful to man. They relish that most destructive of garden pests, the leather-jacket, and, possibly because they tend to be less cautious when

Badger family leaves its subterranean refuge

away from their sett, Eric once found a badger raiding his dustbin: it seems that a hungry boar had learned how to nuzzle off the lid and pick out the titbits of bacon rind, bread and two veg! On mild damp nights earthworms are especially delectable for badgers, as for foxes, and form a large part of their diet particularly in autumn when they spend most of their time feeding in readiness for the rigours of winter. One badger was seen to eat 1083 worms – that is about one-third of its own body-weight – in just 90 minutes, each worm neatly plucked from the earth like an elastic strand of freshly cooked spaghetti. In areas where worms are not plentiful, acorns and berries are an alternative diet. Before each night is out the badgers may pay visits to the latrines that border their territory and mark them with the strong secretion that tells badgers from other territories to keep out.

When it came to the task of filming underground in the sett, that same warning – 'keep out!' – applied as much to Eric Ashby as to all other badgers in the neighbourhood; not so much because of the smell but because of the physical impossibility of filming this most sensitive and private animal in the very heart of badger territory deep underground.

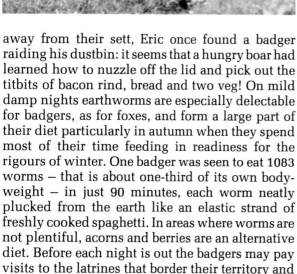

Optimistic to the last, Eric decided to build his own sett to suit his purposes as well as those of a nearby badger family. He collected drainpipes which when concreted together led to two artificial nest chambers inside a woodland shed. With a sheet of glass separating him, his smell and the camera from the two grass-lined nest areas, he hoped that a visiting badger might be persuaded inside to behave naturally beneath the film lights. He planned to increase the power of the lights slowly, night by night, until it was bright enough to film but not a glaring distraction to the temporary tenant. When covered over, all that was visible were the entrance pipes. It was hoped that the mystery of the dark interior would stir a badger's curiosity. Eric sat back, alone in the dark shed, and waited for the action.

It was to be three years before his patient and

OPPOSITE LEFT *Badgers on the way to feeding grounds*

OPPOSITE RIGHT *Scavenging at a dustbin*

OPPOSITE CENTRE *Artificial badger sett under construction*

RIGHT *Sleeping in the sett*

BELOW *The living quarters*

technical ingenuity were fully rewarded. After endless attempts, from now only a metre or so away, Eric became the first human to film badger behaviour actually inside a sett. His observations revealed the cleanliness of the living quarters and the care taken to renew bedding. It showed badgers sleeping peacefully, totally oblivious of their watchful human companion. He photographed the greeting that is exchanged when two badgers meet – the soft purr and the mutual marking with scent. More dramatically, Eric filmed two boars challenging each other for dominance over a female. She would have nothing to do with either of them, and was content to look on as the pair intimidated each other by standing tall to appear all the more formidable: 'They go for each other's tails, get hold with their teeth and shake for all they're worth . . .' This vigorous show of strength was too much for the intruding male. Quickly driven from the chamber, peace was restored. The dominant boar lay back, licked a paw and scratched hard. He had won the game – and sett – but remained blissfully unaware that *their* singles match had been played before the admiring eye of just one spectator inside one of the most original film studios ever built by a wildlife cameraman.

20TH CENTURY FOX   producer Mike Beynon

The vixen is just able to stand in the confines of her underground earth. She has to stand because her four fox cubs are now so big that standing is the only way in which she can accommodate them all at one time. The cubs pummel the vixen's belly as they squabble among themselves, greedy for her milk. It must be uncomfortable, even painful for her: the cubs' teeth are already sharp, and they knead her nipples with their forepaws to encourage the flow of milk. At the entrance to the earth, the dog fox's face appears, and the vixen moves to follow him outside. Sated, the cubs fall off her nipples and, squeaking with excitement, scramble to the entrance and follow their parents outside.

All this has taken place underground, in complete blackness; the foxes themselves cannot see one another. But this vulpine domestic scene has just been witnessed by a million or so television viewers. And they have seen it happening live.

That was part of a four-year affair between the urban fox and the BBC Natural History Unit that was to produce 15 television programmes, and go some way to dispel the popular image of the fox as a solitary vicious killer. But it was a project that had to solve two great problems. How do you film at night – because foxes are largely nocturnal, and how do you film underground – because the vixen would give birth in an underground earth? We decided to tackle the second problem first.

The solution cropped up during a survey of fox earths in Bristol for the Nature Conservancy Council. Waste ground, allotments and church-yards yielded many earths, but, surprisingly, half of all earths found were in the back gardens of private homes, under sheds and garages. Still more

LEFT *Young fox – alert, healthy, coat in perfect condition. The opposite of the thin, mangy villain of folklore*

OPPOSITE ABOVE *Badger meets fox at night. Were they to fight the fox would certainly come off worse*

OPPOSITE BELOW *An overgrown churchyard – the perfect home range for an urban fox*

surprising there were dozens under, or even *in* people's houses.

There is a high proportion of old Victorian properties in Bristol, many with cellars, coal holes, broken ventilation bricks, old-fashioned damp courses: a hundred different ways for the adaptable urban fox to gain access. In one house, old cement around a pipe had crumbled away, leaving a hole through the wall into the cellar. In another house, a missing air brick gave access to the space under the floorboards of a downstairs room. Here a vixen had successfully raised a litter of cubs. It was the discovery of earths such as these that prompted the idea of putting our film equipment inside the cellar of just such a house. We would have the opportunity to observe urban foxes in what was to them a totally natural environment.

The idea was a start, but it created more problems than it solved. It was obvious that glass would have to separate any camera from the foxes themselves. But what about noise, and, above all, what about light? Even the image intensifier that can make a film camera 'see in the dark' requires a small amount of ambient light. And any light at all would obviously make a fox bolt. The solution was to extend a technique pioneered in Britain: infra-red electronic recording. In 'Night of the Fox' (1976) and 'Badgerwatch' (1977) infra-red sensitive Outside Broadcast cameras and infra-red lighting had been used with success.

But this time, instead of for a week or so, the equipment would have to stay in place for almost a year. And instead of waiting outside for badgers to come out of their sett, we would be waiting inside for foxes to come in. And if for any reason they did not come in, a costly project would have failed right at the beginning. We decided to go ahead, anyway. A small team of specialists started adapting the infra-red equipment for our own special purposes. There would be two cameras in the cellar, both operated by remote control from a control room some distance away in the BBC, so no cameraman would be present in the cellar to disturb the foxes.

The next thing was to find the right house, and fortunately for us, the BBC was the owner of just what we wanted: a derelict Victorian house with a garden and, protecting the garden from the road, a high stone wall. Best of all, it had a cellar.

Infra-red light is beyond the visible spectrum for human and fox eyes. The preparations therefore continued with the installation of infra-red lights

in the cellar. To foxes, the place would appear pitch black, but with the installation of specially sensitive electronic cameras, we would be able to obtain perfect television pictures, albeit black-and-white. Finally, microphones were hidden to pick up the sounds made by the foxes.

At this stage a critical decision had to be made. We knew that there was a good chance – better than 50 per cent – that a wild fox would find our specially 'bugged' earth. But with all this expensive equipment installed, could we afford to take the

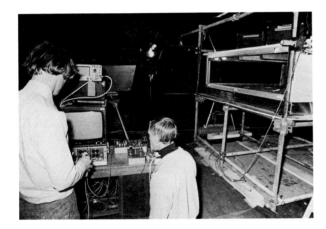

chance? We decided that we must be certain, so a pair of town foxes were brought here specially and released into the garden, which we had made escape-proof. The animals were therefore not totally free-ranging, and the results of our experiment, as applied to town foxes in general, might possibly be called into question. But the animals would have a considerable amount of freedom, and our scientific advisers told us that our observations would be more representative of fox behaviour than any other controlled situation to date.

In October 1978 the vixen and dog fox were installed in the garden and the cameras were switched on. It was the most critical moment of the project. Would the animals find the entrance to the cellar and go in? And once in, would they stay? It was vital that they should feel secure in their earth. If anything – a strange smell or a camera noise – disturbed them, they might have bolted. The vixen might never have returned to the cellar to have her cubs there, and the project would have been doomed. But, as viewers now know they did find and adopt the earth, or the 'Foxwatch' programmes would never have taken place. It was a nail-biting time for all of us as we watched the vixen nervously exploring her new underground home,

ABOVE *Setting up the infra-red equipment which was to stay in place for almost a year*

LEFT *A vixen stuck half out of a broken air grill – way in for her cubs to the cellar of a Victorian house. The vixen was released unharmed*

RIGHT *The vixen comes into the artificial earth for the first time. Will she stay?*

leaving, returning, and finally, after much hesitation, settling down in the corner.

Throughout the winter, our observations continued. Foxes are, of course, traditionally nocturnal animals, so for many long nights we, too, stayed up, together with cameramen, vision engineers, and videotape engineers, for everything was being recorded on to videotape.

The spools of recorded material built up into a sizeable pile as we watched the vixen and the dog fox settling into the garden and earth, coping with the snows and bitterly cold weather and supplementing the food we gave them by catching worms, insects, mice and birds.

In January, when the foxes of Bristol were at their most vocal, our foxes joined in too, making sleep difficult for slumbering Bristolians. In February, they mated, and, on 10 April, the vixen gave birth.

She had dug herself deep into a corner of the earth, so that the moment of birth took place,

ABOVE *While the cubs are still very young, the dog fox has been expelled into the garden*

OPPOSITE *The vixen and cubs in the underground earth*

frustratingly, out of camera vision. But soon the eavesdroppers in their tiny control room heard the first squeakings of the newborn babies and, some time later, caught the first sight of a tiny cub, when the vixen picked it up in her mouth, got up, turned round and settled down again. But there was no way of knowing how many cubs had been born. In fact it was nine days before we knew for certain, nine days when the vixen never left her cubs for one moment, her right flank worn bald by her never-changing position. When she left the earth for just a few moments we counted four tiny heads.

The maternal devotion of the vixen was just one of many fascinating aspects of fox behaviour that we recorded. When they were only a few hours old,

the dog fox, expelled from the earth by the vixen during the last part of her pregnancy, demonstrated a remarkable attentiveness. He fed her continuously for those nine days, appearing at the entrance to the earth, calling to his mate with a specially low, staccato call that signified to her that he was about to enter, and then coming in with food in his mouth and dropping it right by the vixen's side so that she did not even have to get up. Later we saw, too, that the vixen had laboriously stocked up her own personal larder in the cellar. She had 'cached' morsels of food all over the floor of the earth, so, when hungry, she did not have to move more than a foot or so to uncover a titbit.

It had always been my intention that these underground scenes should form part of a larger programme about urban foxes in general. But the pictures we were seeing of the vixen and her cubs were so fascinating and of such good quality, that someone asked why this material should not be put out as a live programme, or better still, a series of live programmes. And that is how 'Foxwatch' started. Late-night viewers on BBC-2 were now able to share with us the intimate scenes of the mother raising her cubs during their first few weeks of life.

During the 13 programmes we saw them covered, at first, in dark hair, with eyes tight shut, then moulting through to their lighter puppy coat and, at 14 days, their eyes opening. We saw their first wobbly movements away from their mother's side and their first meal of solid food – the flesh of a house mouse. We watched as they ventured out into the garden for the first time, saw how they became more playful and adventurous, and, inevitably, more independent.

The whole 'Foxwatch' project, however interesting and successful it might have been, had really side-tracked me from my intention of making a

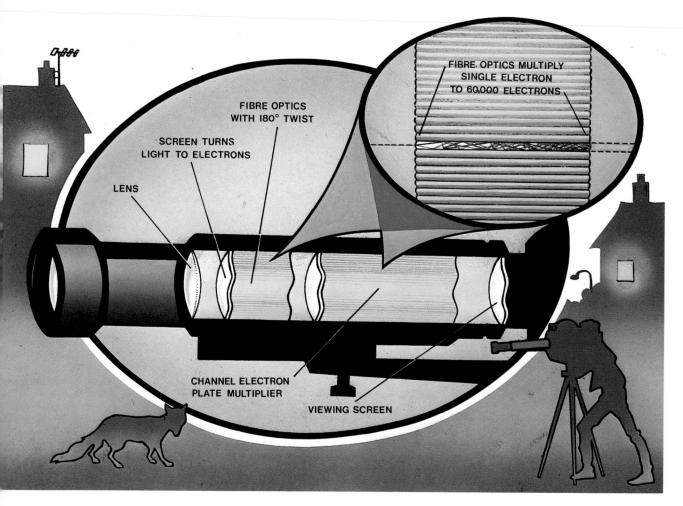

FIBRE OPTICS MULTIPLY
SINGLE ELECTRON
TO 60,000 ELECTRONS

FIBRE OPTICS
WITH 180° TWIST

SCREEN TURNS
LIGHT TO ELECTRONS

LENS

CHANNEL ELECTRON
PLATE MULTIPLIER

VIEWING SCREEN

programme about the wild foxes that inhabit our cities. I wanted it to be a thorough study that showed every important aspect of fox behaviour, and closely examined the creatures' food sources. Do urban foxes really spend all their time scavenging household waste from dustbins? Will they really attack domestic cats? Are they in fact the half-starved, disease-ridden creatures that many people would have us believe? However, it is one thing filming a family of foxes in an enclosure, as we did for 'Foxwatch'; it is quite another thing filming foxes in action in the wild.

This was where our other main item of equipment for filming at night came into its own: the image intensifier. Yet, even when armed with a film camera and image intensifier, how could we be sure where to go? The answer was supplied by a great deal of painstaking research carried out over many years, not by us, but by Dr Stephen Harris of Bristol University. His results showed that Bristol

had the highest recorded population of foxes in the country: in parts of the city up to four families of foxes to the square kilometre (or 10 families to the square mile).

Every year, the 500 adult foxes in Bristol produce 1000 cubs. It is not surprising therefore that most town folk see a fox from time to time at night. What is surprising is that they do not see foxes more often, for every night, from the time the traffic ceases until sunrise, they are on the streets, in search of food.

Far from being half-starved, the truth is that most town foxes are well fed. Where do they get their food? From all the same sources as the country fox, plus a great many more that the town alone supplies. The town fox, like his rural cousin, can hunt for small mammals, fledgling birds, worms and berries, for he is an omnivorous animal and not solely a carnivore. In a notable sequence in the programme, two foxes were seen picking up

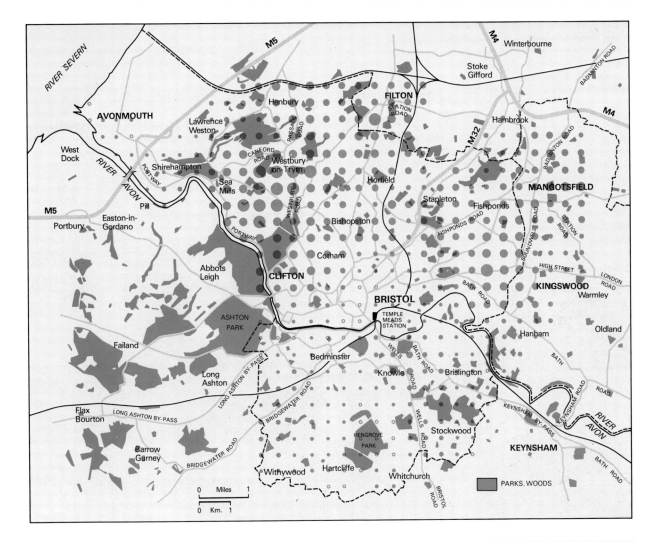

OPPOSITE *The secret of filming at night: the image intensifier amplifies what little light there is and displays the image on a tiny screen. The camera then photographs that screen*

ABOVE *Bristol has the highest recorded population of foxes in Britain. At its densest there are up to 10 families of foxes to 2.59 km² (1 square mile)*

RIGHT *Many a city fox has his diet supplemented by a kindly local resident*

worms off the surface of the grass on a warm wet night in the very heart of a city. It is hunting made easy, for, as we see, one fox can pick up 20 worms in a square yard of grass, and 200 worms provide him with his required food intake for a whole day.

Our long, cold nights filming from hides and parked vans with image intensifiers yielded other fascinating insights into how urban foxes feed; for example, a fox that came to a bird table in a suburban back garden to feast on the leftovers of an evening meal thoughtfully put out for the early-morning birds. That particular sequence also revealed the great agility of the fox in climbing, and even walking along a fence – more like a cat than a member of the dog family. On two separate occasions we filmed foxes 'caching' food: the fox stores food to be eaten later by digging a hole in the ground with its forepaws, dropping the food in, and covering the hole up again with its nose. On one evening our image intensifier, concealed behind the window of an upstairs bedroom, filmed one fox 'caching' food, and a second fox later the same night finding it and digging it up.

The sequence that everybody wanted to see – a wild fox raiding a dustbin – was recorded not by the image intensifier but by the same infra-red equipment that had been used in 'Foxwatch'. Infra-red spotlights lit up the dustbin at night with a light invisible both to humans and foxes. The large infra-red sensitive cameras, moving slowly by remote control, were positioned only some 23 m (25 yd) from a dustbin on one evening and a litter bin on another evening. The latter location produced really spectacular results, with a vixen coming time and time again to raid the same litter

ABOVE *On hot summer days, foxes often go high up on buildings or roofs – seeking coolness and security*

BELOW *Fox raiding a dustbin in the early morning*

bin for scraps of Kentucky Fried Chicken. First she tore off scraps of cardboard hanging over the side, then she jumped on top and foraged inside; finally the vixen got right inside the litter bin, almost disappearing from view as she searched for yet more food to take back to her cubs.

Perhaps the sequence that a lot of people remember is the one showing foxes sleeping in the summer sun up on the rooftops of Bristol: a fox dozing on the roof of a bungalow, another in the guttering on top of a college, on a garden shed, a fire escape, and even a fox sleeping up a tree. How did we get those pictures? With good research, a lot of luck and the help of dozens of local residents. We had sent a circular to many homes in Bristol asking for information on foxes and pleading for a phone call whenever possible. I even went on local radio and television to ask people to phone me at the Natural History Unit if they ever looked out of their window and saw a fox. When the phone call came through, we would simply drop everything, try to find a willing cameraman, and rush round to the address. Most times we arrived too late. Occasionally we got the shot we wanted.

It was as a result of a phone call from a Bristol housewife that, after many unsuccessful attempts, we filmed the final stages of the courtship of a dog fox and vixen. The animals mated time after time in the middle of a small back garden lawn, in the middle of the day, with neighbours digging gardens and hanging out washing all around them.

In a modern world, where wildlife habitats are disappearing and where many species have been pushed into extinction, the urban fox thrives. So do other species. Some seabird populations have expanded – gannets, kittiwakes and herring gulls, for instance. Why? Because of their ability to adapt to living in unnatural environments. If, to a kittiwake, a window ledge represents the equivalent of a cliff edge, so, to a town fox, the underside of a garden shed represents the equivalent of the traditional country earth.

What if he is an opportunist who will eat your strawberries and turn over your dustbin, keep you awake at night and make your garden smell – in short, a confounded nuisance? So what? Does that make him vermin? Is that sufficient grounds for calling in the local council to kill him? Many people would say yes. And who am I to say they are wrong? It is just that, to me, he is a large, beautiful, intelligent wild animal. I cannot hold it against him that he has come to live with me in the city.

After many unsuccessful attempts we filmed the final stages of the courtship of a dog and vixen

2. FURTHER THAN THE EYE CAN SEE

AMOROUS AMPHIBIANS producer Caroline Weaver

As the glaciers retreated at the close of the last Ice Age, lands that had long lain barren under a thick covering of ice became exposed to the sun, and plants and animals gradually migrated northwards to colonize them. Among these animals were amphibians, and a few species made the long journey to Britain, which was then connected to the continent of Europe by dry land. More could probably have established themselves here, but by the time they reached what is now northern France, it was too late. The sea level was rising as the melting glaciers poured millions of gallons of water into the oceans, and once the sea had flooded into the English Channel no more amphibians could reach Britain, since these damp-skinned creatures are killed by immersion in salt water. As a result, there are only six species of amphibian – one frog, two toads, and three types of newt – found in our islands, compared with about sixteen that inhabit northern Europe.

Today, with the growing demand for agricultural land, even these few species of native amphibians are under threat, and their numbers are falling as their breeding sites are destroyed. Although amphibians feed mostly on land, they must return to water, usually a pond, in order to breed. Small ponds, nestling among patchwork fields, were once a common feature of our farmland, but as agriculture becomes more intensive,

and small fields are joined to make larger ones, pockets of unproductive land are no longer tolerated. Ponds, which for centuries have been the spawning grounds of frogs, toads and newts, are drained and filled in with soil. Even where lakes and ponds are allowed to survive, they are often polluted by pesticides which run off from the fields, and these chemicals can kill the amphibians' eggs. Pollution of another kind occurs when roadside ditches are drained into ponds, carrying with them oily residues from the road surface. This forms a film on top of the water and makes the pond uninhabitable from an amphibian's point of view.

A vital refuge for our amphibians is now provided by garden ponds, and it is on just such a pond that two biologists, Tim Halliday (Open University) and Nick Davies (Department of Zoology, Cambridge), are studying their breeding habits, collecting basic information about their behaviour that may help to save these hard-pressed creatures from extinction. Several species breed here, including the common frog *Rana temporaria*, the common toad *Bufo bufo*, and the smooth newt *Triturus vulgaris* and the great crested newt *Triturus cristatus*. Their courtship and egg-laying all take place in a few weeks of frantic activity during March, April, or sometimes early May. Each has a different strategy for ensuring that they produce as many young as possible, and the details of their courtship rituals make an intriguing story, which has taken several years of careful research to unravel.

'A frog he would a wooing go'

The frogs are the first to arrive at the pond, their migrations triggered by a slight rise in temperature and the onset of mild, damp weather suitable for breeding. Their breeding season is so precisely attuned to the weather that it may last for as little as two days, and it is therefore essential that the frogs find a mate as rapidly as possible. The male frogs who have already reached the pond set up a chorus of croaks which may attract others, both male and

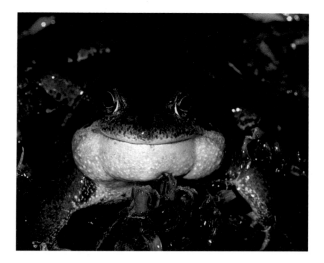

female, and within twenty-four hours several hundred frogs have congregated at the pond. The croaking is vital in ensuring that a large number of frogs assemble at one site, since frogs do not always breed in the same place, nor in an established pond: a temporary pond, a ditch or even a puddle in a rutted farm track may be used.

Why they should need to gather in such large numbers to breed is another question. Biologists call such behaviour 'explosive breeding' and suggest that it may help to increase mating success by making partners easier to find. Another possibility is that the eggs hatch more quickly in large aggregations, since the temperature is higher in the centre of a mass of spawn than in the surrounding water. There may also be a benefit to the tadpoles when a large number hatch together, since the pond's resident predators will quickly become sated and be unable to eat more than a fraction of them. If only a few tadpoles hatched out at once, they would be far more vulnerable. Gregarious breeding has its disadvantages, however, and the males' croaking attracts predators, such as grey herons, for whom the frogs, their attention devoted to courtship, make an easy meal.

When a large number of frogs have reached the

ABOVE *Croaking is vital in ensuring that a large number of frogs assemble at one site*

LEFT *Frogs 'in amplexus'*

OPPOSITE ABOVE *The male develops dark-coloured horny pads*

OPPOSITE BELOW *The female laying eggs with the male fertilizing them simultaneously*

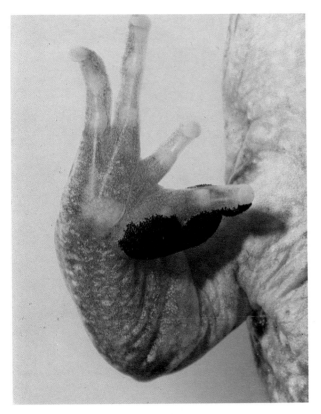

pond, usually within a day of the first arrivals, the chorus of croaks is silenced as the animals retire to an underwater spawning ground. At this stage the animals pair off and each female is firmly gripped by a male, his forelimbs encircling her body tightly so that her egg-filled abdomen bulges out below. The couple are said to be 'in amplexus'. The male develops dark-coloured horny pads on his thumbs during the breeding season, and the thumbs themselves become swollen, to help him keep his grip on the slippery skin of the female. Occasionally a male may grip a female's body so hard that the horny pads break her skin, leaving an open wound.

The ratio of the sexes is usually about equal, so that almost every male eventually finds a partner, but there is a great deal of struggling and scrambling as the males seek out unattached females. Once in amplexus, the male stays firmly attached to his female until she lays her eggs: since he cannot fertilize the eggs inside her, he must shower them with his sperm as soon as they emerge from her body. Up to 3000 eggs are laid by the female, each covered by a gelatinous layer which swells rapidly as it comes into contact with the water, expanding the spawn to about twice the size of the mother within a few seconds. Spawning by one

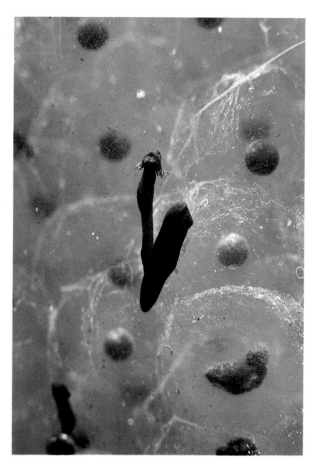

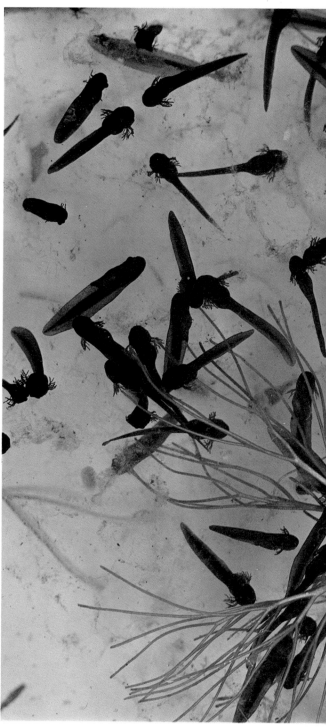

frog triggers off all the others, and within a short time the pond is a mass of spawn. Once the eggs are laid, the males loosen their grip on the females and both leave the pond soon afterwards, scattering into the surrounding countryside and leaving their young to fend for themselves.

The spawn begins to hatch about two weeks later. Even before then some of the eggs will have been eaten by newts, although the gelatinous coating deters most predators. The newly emerged tadpole is less than 25 mm (1 in) long, and has a disc-shaped adhesive organ on its underside which keeps it attached to the jelly for a day or so. It does not eat the jelly, but during this period it draws on reserves of egg yolk which have not already been used up. Gradually its mouth develops, and leaving the jelly it begins to feed on algae. It has a pair of feathery, external gills at this stage, but within three days of hatching these disappear and internal gills develop. (The internal

ABOVE LEFT *Four day-old tadpoles* ABOVE *Frog tadpoles showing feathery, external gills*

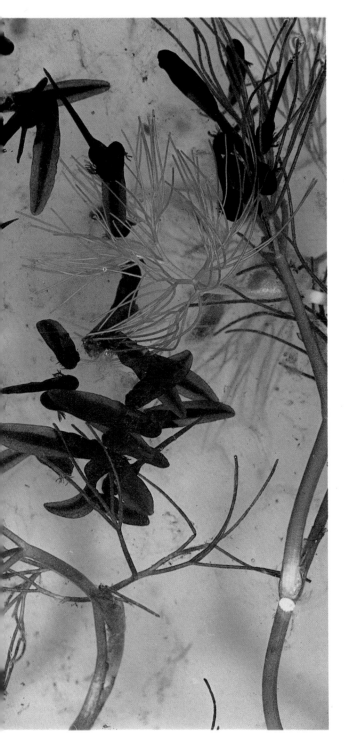

ABOVE RIGHT *Hindlegs begin to develop at seven weeks*

ABOVE CENTRE *The forelimbs appear*

gills will in turn be replaced by lungs as the frog develops into an adult.) As it grows larger the tadpole becomes carnivorous and feeds rapaciously – if an adult frog has the misfortune to die during the mêlée of breeding, its carcass is certain to provide food for the growing tadpoles.

The tadpole has many enemies, especially in established ponds, such as the one Tim Halliday and Nick Davies are studying, where it is likely to be preyed on by fish, newts and dragonfly nymphs (larvae). In ditches and temporary pools there are fewer dangers, but newts may still find them, and so may birds such as crows and coots. Within a few days their numbers are severely depleted and there are only a few thousand tadpoles where once there were several hundred thousand eggs.

Seven weeks after hatching the hindlegs begin to develop and soon each tadpole has a tiny pair of legs kicking away beneath it as it swims. Next the forelimbs appear and the frog's head takes on its

Young male frog

adult appearance with its angular, flattened top and pointed snout. The tail is still present, however, and until it is absorbed a few days later the frog-tadpole is a curious sight, not quite one thing or the other.

While the frogs' eggs have been hatching into tadpoles, another amphibian species, the common (or European) toad, has also visited the pond. Unlike the common frog, the toad only breeds in established ponds with fairly deep water and it is believed to return to the same pond year after year, sometimes travelling several miles to do so. For this reason the toad does not need to set up a chorus of croaking to attract others of its kind, and the croak of the male toad is much softer and less frequent than that of the frog. In this respect the common toad is something of a rarity – most frogs and toads found in other parts of the world behave as our frog does, using whatever breeding sites are available, attracting others to them by croaking.

Although less noisy than the frogs, the toads are far more conspicuous. Males arrive first at the pond and while waiting for the females, they swim or rest at the surface. There is no need for them to conceal themselves from predators, since their skin produces secretions which are extremely distasteful.

The males – several thousand of them – stay at the pond for up to ten days, but the females are there for only two or three days, just long enough to lay their eggs. Only mature females come to the pond, and they are greatly outnumbered by the males, the ratio being as high as five to one for the breeding season as a whole, and far higher at any one time, since the males stay so much longer than the females. The contest between the males for a mate is a desperate one, and it is not unusual to see a struggling ball of toads near the pond's edge, as a dozen or so males fight for possession of a newly arrived female. Occasionally the fight becomes so frenzied that the female is drowned by her suitors.

The instinct to seize a female is such a strong one that male toads respond to almost any moving object of about the right size — a frog, a human finger, the thick stem of a water weed, or even the handle of a fishing net waved in the water. Owners of ornamental ponds that attract toads have occasionally been surprised to find their pet goldfish in the relentless grip of an amorous male toad. The male toad in search of a mate frequently grasps another male, but a croak from his victim, known as a 'release call', tells him that he has made a mistake (the females do not croak), and he immediately lets go.

When Tim Halliday and Nick Davies began their study of toads they assumed that the males who were successful in mating were the larger and stronger ones since they would do best in fights over females, but a series of experiments showed that breeding success was only indirectly related to size. The two biologists began by collecting toads as they arrived at the pond to breed. Some male toads were found to have encountered females during their journey, and were already locked in amplexus when they reached the pond. These pairs were separated, and the male toad measured and fitted with a label before being reunited with his female. The label, a waistband of elastic which was designed to fall off the toad about a week later, was marked with a number that could be discerned from the edge of the pond whenever the toad swam near the surface. The results of the experiment were fairly predictable: as might be expected, small males soon lost their females to other contestants, while large males were able to keep a grip on the females they had claimed. Using captured toads in a tank the same outcome was observed – small males always lost to larger ones. The most interesting observation made during this experiment was that the release call of a small male sounded noticeably different from that of a large male, being higher pitched.

The fact that a male in amplexus with a female

always gave a croak when challenged by another male made the biologists wonder if the croak itself might play a role in the fight. To test this, they recorded the croaks of both small and large toads. Then they silenced a small male by means of an elastic band passed between its jaws and around its forelimbs, which prevented it from expanding the vocal sac on its chest. This male was given a female which he quickly gripped and the pair were then placed in a tank with a larger male. A carefully positioned loudspeaker made it possible to substitute recorded croaks for the male's own croak as he was attacked by the other male. When the croak of a small male was played back, the challenging male succeeded in dislodging him, but when the croak of a large male was played the fight ended

very differently. On hearing the croak of a large male the challenger made only a half-hearted attempt to displace the toad in amplexus and soon gave up when he did not succeed. Clearly, the outcome of the struggle was decided not by the actual size of the toad, but by its apparent size, as judged from the croak it gave.

Tim Halliday and Nick Davies suggest that this may be an advantage to the toads, since it should help to settle disputes rapidly without too much energy being wasted on fighting. When several toads compete for a female at once, however, the system seems to break down amid a symphony of

Toads mating

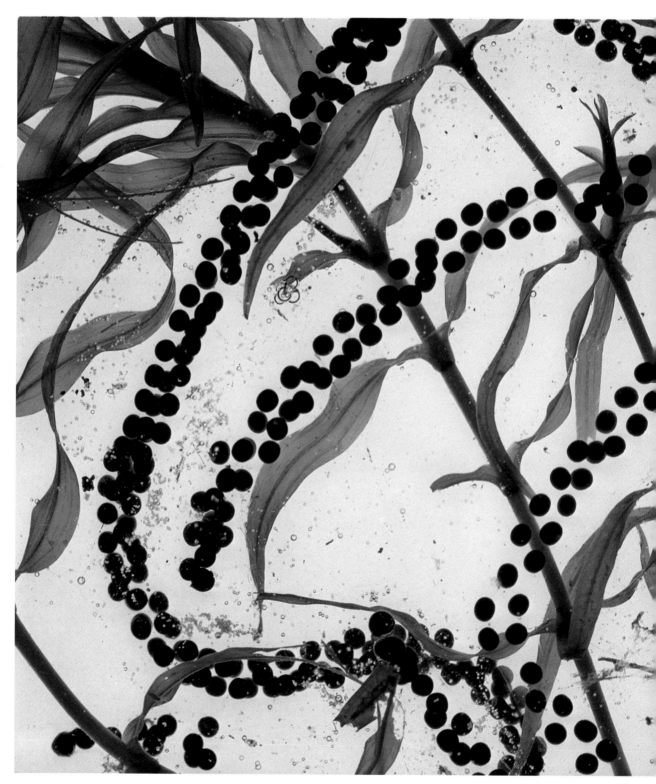

Toad spawn

croaks, resulting in the balls of fighting toads which they have observed in the pond.

Once a male toad has securely claimed a female they retire beneath the water where the eggs are laid. Toads' spawn does not consist of an amorphous mass of eggs like frogs' spawn, but of long, double-stranded threads. The couple swim gently about as spawning takes place and the male kicks with his legs, helping to disentangle the emerging threads of eggs and drape them, like ticker tape, over the water weeds that grow in the pond. At the same time he fertilizes them with his sperm. Some 3000 eggs are laid, in small batches, every half hour.

Once she has laid her eggs the male releases the female and she quickly leaves the pond. The frenzied excitement she aroused in her partner a day or so earlier has given way to total indifference – without her eggs she is no longer attractive to him. Her eggs, meanwhile, are beginning to develop and within a few weeks the tadpoles emerge. They are darker than the frog tadpoles and much smaller, but anatomically they are very similar and their development follows the same pattern. Like their parents, the toad tadpoles are distasteful and predators do not eat them as readily as frog tadpoles.

The young toads and frogs stay in the pond until the summer when, as miniature adults, they scramble out to explore the dry land around them. For the young toads and frogs safety-in-numbers is again the rule, and they leave the pond *en masse* on a day when the weather is auspicious, usually during the month of August.

While the frogs and toads have been engaged in the rough-and-tumble of gregarious breeding, a far less obtrusive, and yet more elaborate, courtship ritual has been taking place in the same pond. For Tim Halliday, the intricate mating dance of the smooth newt has occupied several years of research, and he has finally unravelled the significance of its many separate movements. Long hours watching breeding newts in a tank set up in his Oxford home have given him an understanding of the vital role that these movements play in successful reproduction.

The prelude to mating in the smooth newt is a

change in the appearance of the male, who develops large, dark spots on his back and looks brighter and glossier than before. A wavy-edged crest of skin appears on his back, extending the full length of his tail, and flaps of skin develop around the toes of his hindlegs. The crest of the male is partly decorative, but may be also functional, since it increases his surface area, enabling him to take up more oxygen from the water during his strenuous courtship dance. The colours of the drab yellow-brown female do not change, however, and the male recognizes her largely by her scent.

Having located his female, the male newt embarks on an elaborate and energetic performance to stimulate her interest in him. First he positions himself in front of her, sideways on, so that she gets a good view of his orange belly, his dark glossy spots and his splendid undulating crest. He will follow her about for a while, periodically presenting himself again for her inspection. This is followed by a stage that Tim Halliday calls the 'static display', in which the male uses scent and water movements to arouse the female further. Placing himself face-to-face in front of the female, he whips his tail sharply against the side of his body. This creates a powerful current of water that may quite literally 'knock her off her feet', pushing her backwards, away from him. The male follows this up with a fanning display, in which the tail is curved round against his flank and vibrated vigorously, sending a steady stream of water along his body and towards the female's nose. Tim Halliday believes that this stream of water carries the male's scent – a powerful sexual stimulus – to the female.

After this 'static display' has been repeated for some time the male moves on to the 'retreat display'. This involves the same whipping and fanning movements, but the male moves backwards while performing them, and the female should follow him if all is going well. Once he is assured of her interest the male proceeds to the final stage of his courtship ritual: he stops his tail-fanning, turns around and creeps away in front of her for 10-20 cm (4-8 in). The female should follow, and when the male stops again and quivers his tail he waits expectantly for her to nudge him with her snout. This is the vital signal that makes him deposit a spermatophore (a small elongated sac, full of sperm) and then move forward and turn his body round so that it blocks the female's progress.

She moves forward to this point, which should be exactly one body length ahead of where she stood when the male dropped his spermatophore, four or five seconds beforehand. This positions her cloaca (reproductive opening) exactly above the spermatophore which sticks to the lips of the opening, allowing the sperm to be drawn up inside her body. This tricky operation is helped by a fluid-filled cavity within the spermatophore, which swells up after the male deposits in it, pushing one end of the spermatophore up towards the female's belly. The transfer of the spermatophore is not always successful, however, and as many as 50 per cent fail to be picked up by the female. For this reason, the male repeats the latter part of his courtship dance two, three or even more times, producing a fresh spermatophore on each occasion.

To establish exactly what reactions the male needed to elicit from the female during the courtship dance, Tim Halliday looked for a means by which he could control the female's movements without interfering with the male's responses to her. With this in mind he devised a girdle made of flexible plastic which fitted neatly around the female's waist. The girdle was attached at the back to a long stick so that she could be manipulated from above almost like a puppet. Using this 'strait-

OPPOSITE *Male great crested newt*

RIGHT *Male smooth newt prior to mating*

BELOW *Smooth newt courtship ritual*

jacket' as he calls it, Tim Halliday found that if the female did not give the appropriate response the male would generally break off the courtship dance and go back to the previous stage. The most critical action required of the female was the nudge from her snout which led to the spermatophore being desposited – the male never released his package of sperm until this signal was given.

The business of transferring spermatophores from male to female may seem unnecessarily arduous and complicated, but it gives newts an important advantage over toads and frogs, who fertilize their eggs externally. Whereas a frog or toad must perform mating and egg-laying at exactly the same time to ensure fertilization, in the newt they are two separate procedures. Once the female has absorbed the male's sperm she can store it until the right time for egg-laying arrives. This means that mating does not have to be synchronized with the onset of favourable conditions for egg-laying.

The smooth newt lays about 100 eggs, far fewer than the frog or toad, and each is laid individually over a period of several days. The female uses her hind feet to wrap each one in the leaf of a water

plant, and a sticky secretion from her cloaca holds the folded leaf together. The emergence of the newt tadpoles in late spring coincides with the hatching of daphnia (water fleas), and these tiny creatures keep the tadpoles well fed for several weeks. Newt tadpoles differ from those of the frog or toad in a number of ways. They are equipped with a pair of

OPPOSITE ABOVE *Smooth newt egglaying*

OPPOSITE BELOW *Egg of a smooth newt developing*

LEFT *Smooth newt at the end of the breeding season*

BELOW *Toad crossing a road*

balancing organs and their feathery external gills are not lost until the tadpole metamorphoses into an adult. Unlike frog and toad tadpoles it is the forelimbs that develop first, and the tail, instead of being resorbed, is retained into adulthood.

Of all our amphibians, the smooth newt is the one that is least endangered by the advance of agriculture. It has proved surprisingly adaptable, taking to alternative breeding sites, particularly garden ponds, with alacrity. The outlook for the other two newt species is far less hopeful and one, the crested newt, *Triturus cristatus,* is declining at an alarming rate. The frog and toad are also disappearing, and a survey carried out in 1970 showed that large areas of arable land had no frogs at all, whereas once they were common everywhere. The toad's faithfulness to its breeding site is a handicap in more ways than one for its migrations towards the time-honoured site may today take it across a busy road. The toads' habit of stopping frequently to sniff the air during their migration means that it may take between 10 and 15 minutes for them to cross the road, and during this time large numbers fall victim to the wheels of passing cars. Conservationists in many parts of

Britain now organize 'toad watches' to alleviate this problem. They remain on alert every night during the breeding season, and when the toads begin to migrate they congregate at their usual crossing place. Waiting on one side of the road they collect the arriving toads in plastic buckets and then carry them across to the other side where they are released. Another approach to the same problem has been adopted on the Continent: small tunnels have been built beneath roads where the toads cross, and fences erected to direct them away from the road and into the tunnel.

In the long term, however, it is the filling in of ponds, and their contamination with pesticides or oil that must be tackled, if Britain's amphibians are not to disappear entirely. What we have to decide is whether agricultural production must be increased at all costs, even at the risk of destroying the richness and variety of our wildlife, or whether there is room in Britain for animals, as well as human beings. It would be a shame if those pioneering amphibian species who made the long, arduous journey to these islands at the close of the Ice Age should now be eradicated from them for ever by man. L.G.

THE WATER WALKERS producer Pelham Aldrich-Blake

For most small creatures, the water surface is a trap. If they break the thin elastic layer that covers it, they are spread-eagled and held fast by surface tension. But a few have mastered its peculiar problems, and for them, this fragile interface is home. The programme 'The Water Walkers' examined this miniature world – a world peopled by pondskaters and swamp spiders, springtails and whirligigs, where water bends and the laws of physics take on new dimensions.

Filming such diminutive forms of life is an esoteric and highly skilled craft. Most of the specialized macrophotography in the programme was the work of a small freelance company,

London Scientific Films, based at that time in a fourth-floor office in Great Portland Street. Although some sequences were filmed in the wild, for others the more controlled conditions attainable in the studio were esential. For over a year, it was home for tanks of assorted insects and artificial sets furnished and vegetated with rocks and plants from the natural habitat.

The stars of the show were the pondskaters. These insects dimpling and jerking their way across the face of a pool are a familiar sight, but have you ever stopped to wonder how they can walk on water?

Part of the answer lies in the way that water

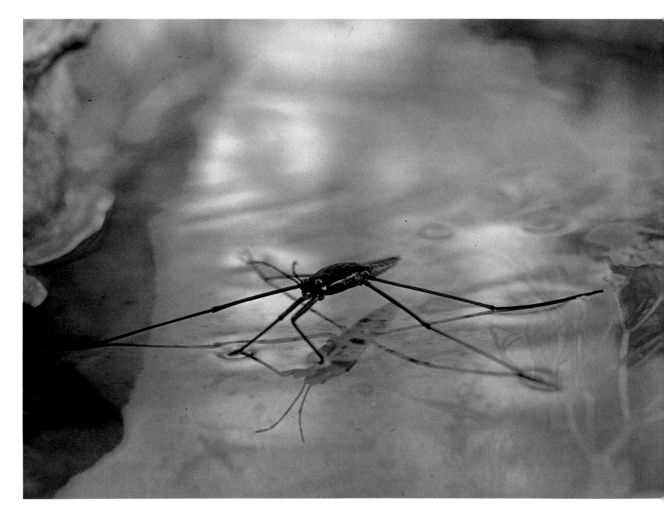

molecules attract one another. Within a body of liquid each molecule is attracted to those around it, and since they are fairly evenly distributed, all these forces cancel out. At the surface, however, it is different. There is still a strong pull on each molecule from those beside and below it, but only a very weak attraction from the much sparser molecules in the air above. As a result of these unequal forces the surface of the liquid tries to contract, and this 'surface tension' forms an elastic layer strong enough to bear the weight of a small insect without breaking. (For the same reason – the action of these unequal forces – droplets of water tend to be round as this shape gives them the least possible surface area.) Surface tension, the problems it poses for small insects, and now they circumvent and exploit it, were the main themes of the programme.

In the case of the pondskaters, however, surface tension is not the whole answer. Other insects may be just as light, but they are often unable to walk on the surface. If you look closely at a pondskater, you will see that only its feet touch the water. The feet are covered with tiny, water-repellent scales, so they just dent the surface rather than break it. The body too has a waxy, water-repellent covering, so even if it does touch the surface the pondskater is not trapped.

The importance of being water-repellent, or hydrophobic, can be seen if we compare an insect that lacks this special adaptation. If it lands on the water it pierces the surface layer much more readily, and once it does so it has little chance of breaking free. It is held down by the pull of surface tension all around it, snared by the same force that keeps the pondskater up. Indeed, creatures that have been trapped in this way make up a high proportion of the pondskater's diet.

Although the bodies of all specialized surface dwellers are at least partly water-repellent, some are a subtle mix of hydrophobic and wettable, or hydrophilic, structures. Springtails, for example, are tiny, primitive insects little bigger than a pinhead. Most kinds live in soil or decaying vegetation, but some are adapted to a life on the water. They are so small that they have to have a means of anchoring themselves to the surface; otherwise

ABOVE *Only the pondskater's feet touch the water*
RIGHT *Pondskaters feeding*

OPPOSITE *The stars of the show were the pondskaters*

they would be blown away by the slightest breath of air. Underneath their bodies they have a short stump, and the end of this is wettable. The water clings to it and holds the springtail down. The rest of the creature's body is water-repellent, and so are its legs, so it does not sink in too deep. Its feet, however, are sharp-clawed and wettable, allowing it to get a grip on the surface and walk.

One of the essentials in keeping and filming any surface-dwelling insect is that the water should be perfectly clean. If there a trace of contamination the surface tension is lowered and no longer has the strength to support an insect's weight. Pond-skaters are becoming much scarcer in the wild for this very reason; pollution with detergents makes them fall through the surface and drown.

This same phenomenon is turned to good use by a creature that possesses the most striking adaptation to moving on the surface, the camphor beetle. This small beetle, only a few millimetres long, lives in vegetation on the edge of ponds and streams. Every now and then it falls into the water, and when it does it gets back to the bank in a most surprising way. From its tail, it secretes a substance into the water that has the effect of lowering the surface tension behind it. It is thus pulled forward at enormous speed by the unchanged surface tension in front, without having to exert any effort at all. (The same principle, using a fragment of mothball or a sliver of soap, can be used to propel a child's toy boat.)

We had hoped to include a sequence on camphor beetles in the programme, but the beetles had other ideas. They performed perfectly when we first dropped them on to the water – but as soon as we tried to film them they decided to walk home instead. On the rare and unpredictable occasions when they did do the right thing, they moved so fast that the camera could not keep up with them. We took several hundred feet of film, enough for a minor epic on the camphor beetle, but despite our persisting right up to the last possible moment when the programme was actually being edited, we failed to get the one vital shot that would make it usable.

With other sequences we had better fortune. One of the most memorable, and one that best illustrates the strength of surface tension, was of pond-skater egg-laying. Sometimes eggs are deposited in the splash zone on rocks projecting from the water, but more often they are laid beneath the surface. To get there, the female has to break through the surface film. It is quite a struggle; her waxy body repels the water, and so long as she merely distorts rather than pierces the surface, the surface tension pulls her back up. When she finally forces her way through, a silvery film of air clings to her body. This allows her to breathe while she is under water, but it also makes her very buoyant, so she has to hold tight to the rocks while she is laying. Getting back to the surface is easier; she just lets go and floats up.

After about 20 days, depending on temperature, the eggs begin to hatch. First the head emerges,

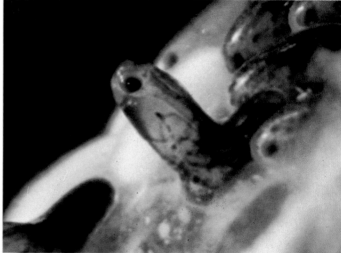

OPPOSITE LEFT *A spring tale on the water surface*

OPPOSITE RIGHT *Female pondskater breaks the surface film to lay her eggs*

ABOVE LEFT *Pondskater eggs*

ABOVE RIGHT *Pondskater nymph emerging from the egg*

RIGHT *Mosquito larvae hanging beneath the surface*

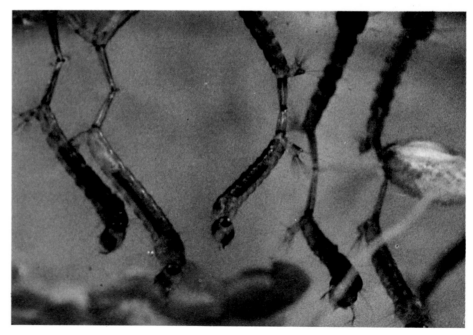

then the body, and finally the impossibly long legs unravel themselves from the confines of the egg. This, too, was captured on film, but only after much patient waiting to be there at precisely the right moment.

Keeping insects alive in the studio for a few days is one thing; maintaining them for months and inducing them to breed is quite another. We were fortunate in having the assistance of Edward Eastwood, a biologist who specializes in pond-

skaters. He was already familiar with the problems of keeping them in the lab, and had devised special tanks to simulate the flowing water of their native streams. The success of 'The Water Walkers' owes much to his advice.

Pondskaters are not the only insects that have sometimes to face the problem of breaking the surface film. The same difficulty affects small creatures such as mosquito larvae that live beneath the water but are dependent on the air above to

BELOW *Mosquito breathing tubes from above*

BOTTOM *Water boatman inverted beneath the surface*

breathe. They are only a few millimetres long, so to rupture the surface layer by brute force would be beyond them. They solve the problem by having microscopic oily flaps at the tip of their breathing tube. These are water-repellent, so when they touch the surface they break through, and the surface tension then pulls the flaps out like the petals of a flower. In fact, this anchors the larvae so firmly that they have to give a convulsive jerk to break free. Once below the surface again, the flaps close over the end of the breathing tube to prevent water from entering.

Water-repellent structures of this kind are common to many aquatic insects dependent on atmospheric oxygen. The water boatman, for example, has a hairy, water-repellent fringe to its tail allowing it to hang just beneath the surface to replenish its air supply, an envelope of gas thus trapped in minute hair-covered grooves on its body.

Many creatures exploit the properties of the

water surface in another way too. They use the ripples generated by struggling insects trapped on the surface to guide them towards food. Pond-skaters, for instance, have vibration detectors in their feet. The order in which these are stimulated tells them which direction the waves are coming from, and allows them to home in on their prey until they are close enough to see it.

The swamp spider, Britain's largest spider, hunts in the same way; instead of a web, it uses the water surface as a snare. It lives in small pools in fen and heathland, often lurking at the water's edge – with at least one of its eight feet on the water to pick up the vibrations that will send it running over the surface to its prey.

The use of surface waves reaches its greatest sophistication in the whirligig beetles. Not only do they detect the ripples made by struggling prey, but they use the wavelets created by their own movements as a kind of aquatic echolocation. They pick up the returning waves reflected from objects in their path, and so avoid bumping into them.

RIGHT *Whirligig beetles*

BELOW *Swamp spider*

FLOWER FROM THE FLAMES producer Caroline Weaver

The variety in the size and shape of the protea flower was so great that Linnaeus named them *Protea*, after the Greek god who could change himself into an infinite number of shapes to avoid being recognized. The shrubs and bushes he described came from the coasts and hillsides of southern Africa, where they flaunted some of the most beautiful and bizarre flowers in the whole continent. Today, botanists have identified 135 species of proteas in Africa, and of these 69 are found only around the Cape of Good Hope. Proteas are so typical of this area that one of them, *Protea cynaroides*, has been adopted as South Africa's national flower. Its brilliant red or pink blooms are covered in fine hairs which give them a silvery appearance, and they may grow to be nearly 30 cm (12 in) across.

The history of these remarkable plants began thousands of years ago, not in southern Africa, but further north around the equator. They gradually spread southwards into the well-watered highlands of the Rift Valley, and from there eventually moved out of the tropics to reach the Cape. In the Mediterranean-like climate of this region, the proteas found an ideal habitat and they flourished to an unprecedented extent, evolving the strange and diverse flowerheads that make them so intriguing.

Each protea flower is in fact made up of a multitude of separate flowers, or florets, all crammed together in a single bloom. The florets' petals are long and thin, and the distinctive colour of the

OPPOSITE Protea cynaroides
BELOW *The highlands of the southern West Cape*

50

Individual Floret

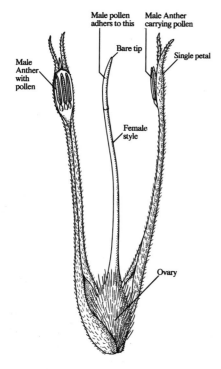

Male Anther with pollen

Male pollen adheres to this

Bare tip

Male Anther carrying pollen

Single petal

Female style

Ovary

Half section through a *Protea* flower-head inflorescence

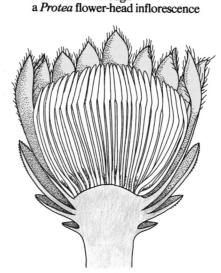

flowerhead is provided by a ring of leathery bracts that surrounds the florets. Bracts are modified leaves, and in most plants they are green in colour and simply protect the flower buds. But in the proteas, the bracts are often a flamboyant red, pink or yellow, and they can assume a variety of shapes. Many proteas have spiky bracts that give their open flowerheads the appearance of brightly coloured, giant artichokes. The paler bracts of *Protea nerifolia*, however, never open wide, but remain folded over the florets like a shroud. They are tipped with luxuriant black tufts, which contrast strongly with the remainder of the bracts, giving the flowerhead a striking appearance.

Proteas flower between May and October, during the South African winter, a time of heavy and persistent rain when there are few other flowers to be seen. Each species has its own way of ensuring that its flowers are pollinated, and the shape of each flower has evolved to attract a specific type of pollinator. Most protea flowers are insect-pollinated, and they make access easy for their flying visitors by opening wide to expose their florets. It takes place when each floret's male

pollen is ripe to be transported to another flower, and before its own female part (the style) is ready for pollination.

Each protea floret consists of a fine tube of petals that surrounds the thread-like style. The dark anthers (male structures that produce the pollen) are attached to the inside of the petals. They start to produce pollen before the flowerhead opens, and this adheres to the style, leaving just the very tip bare. As the style grows, the petals shrivel and curl back into the depths of the flowerhead, and all that is left projecting out of the flowerhead is the style, complete with both male and female constituents for fertilization.

The brightly coloured bracts unfold as the flowerhead opens, and reveal the florets ready to dispense their pollen. Worker bees are frequent visitors, and a single bee may visit five hundred flowerheads in a day, just to collect together a pinhead-sized droplet of nectar. This rich sugary solution is an ideal food for insects, and the plants produce it as a means of luring pollinators to their flowers. Bees fortunate enough to find a *Protea repens* plant in flower can collect a day's nectar at one visit. This protea, known as the sugarbush, produces so much nectar that it oozes from its red flower buds even before they have opened. As a bee moves among the florets of a protea flower, it becomes dusted with pollen, and when the bee visits another flower of the same species, some of the transported pollen will stick to the bare tips of the florets, so that fertilization takes place.

Pollen is also a rich source of food, and bees strip it from the florets and deftly pack it into paired depressions on their back legs known as pollen baskets. Sometimes they go even further than this and rob the plant by breaking open immature florets. Working hard with their mandibles, they chew through the petals to reach the pollen.

One of the greediest plunderers of pollen and nectar is the 32 mm (1¼ in) green protea beetle, which fills the air around protea flowers with the noise of its heavy buzzing. Each beetle spends the first two years of its life as a pale worm-like larva deep in a termite mound, and then emerges as an adult beetle in autumn, just as the proteas begin to bloom. In their eagerness to reach the florets, the beetles break into the flower buds, parting the tough bracts to work their way inside. Sometimes over half a dozen beetles will squeeze into a single flower bud, and although they transfer pollen from flower to flower on their bodies, it would seem that they are so destructive that their value to the protea is insignificant. After they have visited a flower, they leave a mass of broken florets, only some of which will be able to set seed.

A large protea flower offers food and concealment for a whole community of small creatures. In just one flowerhead, over one thousand animals have been counted, including some species that were until recently unknown to science. The green protea beetle is a giant among these flower-dwellers, and it acts as an unwitting carrier to one of the commonest kinds of animals found among the florets – tiny mites. A close look at a protea flower will sometimes reveal a seething mass of these mites, most of which scavenge for dead organic matter or pollen. Because they have no wings, they are unable to get from flower to flower without assistance, and this is where the green protea beetle comes in. When a visiting beetle prepares to leave a flower, it spends some time flexing its muscles to raise its body temperature. (A heavy insect like a beetle uses a lot of power in flight, and its wing muscles can only work best when its body is warm: by flexing its muscles and taking in oxygen, the beetle can raise its temperature by as much as 12° C [54° F], enabling it to take off in search of another flower.) The movements of the beetles have a dramatic effect on the mites. Suddenly, they cease their random wandering among the florets and rush towards the beetle, scuttling over its body and then congregating under its chin, away from the beetle's flailing legs. When the beetles flies away, it is forced to take the mites with it. On landing the mites disembark and find a new source of food, perhaps pollinating some of the florets in the process.

Insects are not the only animals that sometimes feed on protea nectar without helping in pollination. Small iridescent sunbirds, which are similar to American hummingbirds, sip at the nectar with their long tubular beaks, but then move on without coming into contact with the pollen at all. There are many species of sunbird in southern Africa, and they feed at a wide variety of flowers. Some plants have evolved ways of making use of them as pollinators, but not the proteas, and for them the sunbirds' visits go unexploited.

Yet there is one bird that *is* a very effective

pollinator, and this species, the Cape sugarbird, is found only where the proteas bloom. The male is a buff-coloured bird with an elegant tail that makes up two-thirds of his 430 mm (17 in) length; the male Cape sugarbird is often seen sitting at the top of a protea bush, sometimes in the company of his less colourful, shorter-tailed mate. Sugarbirds feed on nectar and insects, and since few other plants flower during the winter, they feed almost exclusively on proteas when they are in bloom, drinking their nectar and catching insects attracted to the blooms. The birds do feed on all proteas especially species like *Protea nerifolia*, which has deep cup-shaped flowers offering an abundance of nectar. As the sugarbird lands on a flower its feet spring open the florets, and the bird's forehead becomes dusted with yellowish-red pollen as it pushes its beak deep into the flowerhead for nectar. When the bird moves on to the next flower of the same species – it visits about 250 separate flowers every day – it spreads pollen onto the exposed florets. The sugarbirds and the proteas have evolved a high degree of dependence on each other. The bird gains a reliable source of food and the plant an effective pollinator which, unlike many others, spreads the pollen without eating it as well.

The flowering of the proteas is so important to

LEFT *Sunbird*

RIGHT *Cape sugarbird on*
Protea eximia

the sugarbirds that they raise their young only when the blooms are open. The male selects a territory where he has exclusive rights to a good supply of flowers, and he perches on the plants he has claimed to ward off other males by singing. As the breeding season begins, he tumbles and flutters in the air to attract a mate, rising and falling over the proteas. When a female has accepted him as her mate, she makes a cup-shaped nest deep in a protea bush, shielded from predators and the cold winter winds. The proteas provide building material for the nest, and fluffy seeds form a soft lining on its inside. When the nest is nearing completion, she feeds mainly on insects. They are either caught on flowers or snapped out of the air in a short flapping flight by the hungry bird. Nectar is a very energy-rich food, and is ideal for the male who must defend the territory, but to produce eggs the female needs protein as well, and insects fulfil this requirement.

During an incubation period of 17 days, the female bird leaves the nest for short feeding sorties about every two hours. The male makes no attempt to feed his mate, but instead spends his time patrolling the territory and preening his feathers. Once the chicks have hatched out, more food is needed, and the female begins collecting small insects, spiders and nectar. At this stage the male occasionally joins in and helps. By the time the nestlings are 12 days old, the female is spending all the hours of daylight feeding them, and a continual supply of cockroaches, flies, bees and ants is brought to the nest. By the time the chicks are 18 or 19 days old they fly from the nest, and within the space of a further three weeks the parents' urge to protect their young disappears, and the young birds are chased away to find protea clumps of their own.

Until recently, the life of the Cape sugarbird was thought to show one of the most complex inter-relationships between plant and pollinator found in the proteas. But in the highlands of the south-west Cape, an even more unusual story has been unravelled. On these rocky hillsides, dense bushes of *Protea humiflora* have flowers so hidden in their tangled branches that it is unlikely that they would attract any insect or bird. In addition to this, not only do the flowers hang down facing the ground, but open only at night. The answer to how these flowers were pollinated finally emerged a few years ago when two scientists decided to look beyond the flowers themselves.

Occasionally, a little way from a protea bush, they found a seed head, or a pile of seeds, some of which had been chewed. By looking carefully over the ground, they discovered that one flowering plant would be linked to another by barely discernible runs, sometimes dotted with droppings. Evidently the heavy pendulous flowers of *Protea humiflora* were attracting nocturnal visitors, so after dark the scientists kept watch. In the still of the South African night, small rodents, most commonly the Namaqua rock mouse, were seen scuttling along the pathways and climbing up into the protea bushes. Guided by the flowers' yeasty scent, they clambered over the florets, sipping the nectar which the flowers produce at night, and chewing at the fleshy bracts. For the mice, like the sugarbirds, the protea flowers are a valuable source of food at a lean time of year. As they visit the flowerheads, their noses become dusted with pollen which they spread from bush to bush – one of the few examples of pollination by rodents known in the world.

When pollination is complete and seeds are produced, most proteas shed them as quickly as other plants. Yet when the florets of *Protea repens* have been pollinated, the bracts close up again, and remain so tightly closed that seed cases may stay on the shrub for up to 20 years. At first sight this seems a lost opportunity to reproduce, but in fact the seeds are not wasted, they are merely waiting. For year after year the protea may grow and the seedcases stay intact, but then one fierce acres of proteas are destroyed as a fierce bushfire sweeps through the dry vegetation. The seedcases are not burned, however, but just charred, and the bracts which once clamped the seeds together curl back and turn to ash. Deprived of sap on the black, lifeless bushes, the seedcases fall to the ground and open to expose the seeds. As the wind blows it catches the seeds' feathery parachutes and scatters them. In the rains of the following autumn they will germinate, springing to life after years of dormancy. For many protea species, this phoenix-like rebirth from the flames of a bushfire is just one more chapter in a life story full of surprises. Within a few years the proteas are growing again, and the countryside is ablaze with their colour and alive with the movements of the animals that live so closely with them.

L.G.

Namaqua rock mouse

LADY OF THE SPIDERS/
ENCOUNTER UNDERGROUND *producer Peter Bale*

There is a click as sprockets lock the roll of film into place on the editing machine. Chairs shuffle as the three of us choose our place in front of the tiny viewing screen. The main lights cut out and our first viewing session begins. There will be six hours of this before a foot of film is chosen for a place in the twelve or so sequences that go to make up each complete story for 'Wildlife on One'. With me is Dione Gilmore, my co-producer, and Scott McClennan. It is Scott's first viewing of material that has taken months to film and many more to plan. As Film Editor he is making his first clinical assessment. Dione and I are already familiar with the detail of our production so we need to know how wide of the mark our judgment has been and whether or not we actually have our story on film.

'SPIDER LADY' *Location* Bungalla
             Producer Peter Bale
             Camera Jim Frazier ROLL ONE

These words are scrawled on the back of a postcard rather like a hastily written luggage label; filmed by the cameraman at the start of every new magazine, the words identify each roll, making sure that it is allocated to the correct production. Scott McClennan ticks off the roll number on his shot list. Lifting his mug of Australian tea he settles down to watch. All eyes are on the screen.

The making of wildlife programmes is very much an international affair and BBC staff travel worldwide in search of good subjects and stories. So it is that a producer from Bristol, England, comes to be working on a co-production made in Western Australia and now being edited in Melbourne, with a Director and Editor from the Australian Broadcasting Commission. We have a common interest in producing the best possible programme; soon audiences as far removed as Perth and Preston, Sydney and Sacramento will be watching the results of our joint enterprise about the private life of one of the world's most ancient inhabitants: a spider!

It has always been something of a mystery to me that among the small wild creatures, insects and spiders tend to be regarded with a suspicion that verges on the frenetic; it may be their tendency to creep and crawl or to appear before us suddenly and so unexpectedly. Such phobias, witnessed among friends and family when confronted by even the most innocent 'eight-legger' found lurking in some unlikely corner, suggest that to be small is not necessarily to be beautiful. Bearing in mind the diminutive proportions of the antagonist, it is rather strange that so many of us should feel so much, so strongly, about so very little. Of course there are a lot of them – countless more than human beings – but it is not by sheer weight of numbers that spiders can drive people to the nearest high chair – one can do the trick!

This human reaction – or weakness perhaps – has been reflected in the attitude of many TV viewers ever since the natural history programmes took to the screen. We are told that programmes about furry animals are the most acceptable, that snakes are far from popular, and that spiders and the like are rated big 'switch-offs'.

Television producers must be aware of their audiences and their likes and dislikes, but an irrational audience reaction can raise problems for the producer wishing to do justice to all the subjects within his range. Should he reject a subject just because 'Aunt Mabel cannot abide ants or spiders, or snakes or alligators? Of course the reply has to be 'Certainly not!' By all means let us be sensitive to viewer wishes but not at the cost of denying television exposure to all spiders and ants wherever they may live. They share our world, they have a considerable effect upon it and upon us humans, so we would be wise to find out as much as possible about their behaviour. It is impossible not to be impressed by the vast range of species that occupy the nooks and crannies of our planet. Many of them were around with the dinosaurs. Some managed to survive and multiply with devastating success, so that today their multitudinous presence commands attention. I had been reliably informed that if I cared to look close enough for long enough, I would be amazed at the behaviour of one particular Australian spider: *Anidiops*

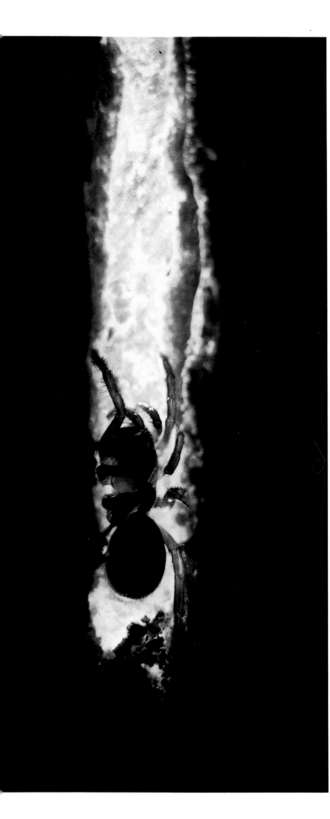

villosus. But would it make a good television story? That is how I came to be heading east from Perth on that hot late summer day.

The car headed on at a steady high speed, the windows closed to keep out the flies. The brilliant blue of the Western Australian sky stretched out endlessly on all sides, with only a trail of brown-red dust rising from our wheels to mar the landscape or give notice that humans were about in this stunning country. My companion and driver was the very same 'Spider Lady' that cameraman Jim Frazier would identify some twelve months from now at the start of each new roll of film.

This working title was chosen not for the obvious link with the popular American series 'Spiderman', but for immediate recognition, for that is exactly what Barbara York Main was: a 'Spider Lady'. The title appalled Barbara. No respectable scientist would wish to be labelled in such a manner! I was able to reassure her that we had no intention of using the offending words as a title; perhaps 'The Lady of the Spiders' would be more acceptable, and would not lose too much of the intrigue value so essential when presenting the work of a specialized field naturalist to a wide popular audience.

We had turned off the main Kalgoorlie road several miles back and were now heading north to the small wildlife reserve that Barbara had created just off the road to Bungalla. This was Australia as it used to be before the bush had been slashed and burned to make way for the countless miles of wheatfields that now lie as far as the eye can see. Swinging off the dirt road onto an even less stable track, we jerked to a halt among tussocks of grass, grey bushes, anthills and termite mounds. As I opened the car door and exposed my eyes, ears and mouth to the sweat-seeking flies, it struck me as being a most undesirable place. Devoid of bird song, with no breeze to stir the acacia bushes or waft a cool draught among the gnarled, twisted sheoke trees, it stopped me dead in my tracks. I found it hard to believe that these few sun-baked square yards could be such a source of wealth to a scientist with an international reputation. What was here that could possibly be worth revealing?

Pulling her floppy cotton hat tight down over

Young trapdoor spider in excavated burrow

her bunched hair and shouldering a small canvas bag full of lethal-looking instruments — more reminiscent of those found in a dental surgery — Barbara beckoned me to follow her into the bush. This 'Pom' followed, sporting a pair of shorts. Wasn't this how Aussies dressed when bush-whacking? Jagged twigs began to scrape bare knees and graze chunks of leg, but being British to the last drop of blood, I just pressed on through the low undergrowth, swatting flies and ducking under the canopies of casurinas in the wake of a lady more sensibly dressed — protected by tough cord slacks from the blood-letting attacks that I had reason to curse.

At first encounter, the wodjil (the aboriginal name for this brittle, dry bush) appeared monotonous and unproductive. It was no surprise to learn that Barbara's father, like so many early settlers, had set to and hacked it back. Indeed the dream of new golden harvests to come must have seemed infinitely preferable and more profitable. Barbara and her three brothers had been raised nearby on the family homestead. They had grown up out there among the clusters of acacia and

OPPOSITE ABOVE *Densey Clyne looking for spiders at the study site at Tammin*

OPPOSITE BELOW *Barbara York Main*

RIGHT *Plugged nest*

BELOW RIGHT *Fanlike twigs and phyllodes*

casurinas. Very soon this young schoolgirl had come to realize that the wodjil had a special fascination. This attraction was not shared by her brothers – they had better things to occupy them on the ranch – but for Barbara a whole new world was waiting to be discovered under the wind-blown debris of the bushland.

Barbara rounded an anthill, paused, then drew out a glinting knife and began passing it over the leaves at our feet, appearing to cast some magic spell over the litter of random leaves. As if invoking some mathematical wizardry, she began to count; first in single digits, then in multiples of two. At 24 the sharp point of the blade came to rest over one particular cluster that covered a space easily contained within the rim of a small teacup. Barbara squatted and drew my attention downwards: this was no random or accidental grouping. The pattern of leaves was regular. On closer inspection they were in a group that was precise: the arrangement of twigs and phyllodes was fanlike – as if someone had woven a fan of leaves around the edge of an oval sand-coloured disc in the surface of the ground. Some discs were as large as a 50-cent or 50-pence piece, others, the size of a child's chocolate drop. There were dozens of them. As the eye became accustomed to the geometry of each miniature structure there was no mistaking the urge to lean forward gently and open the trapdoors with one of Barbara's scalpel-like instruments. I held back from the brink, so to speak; but,

fascinated, I made the commitment there and then that we would make our next wildlife film about the life of this particular species of trapdoor spider. All that remained to be done was to find ways to film inside the underground homes. What was more, it had to be done right here in the middle of the wodjil.

To film the minute details of a creature that lives just over 60 cm (2 ft) underground and is sensitive to the slightest temperature change even from a low-powered light bulb, is, to put it mildly, difficult! Apart from the special problems of limited depth of field and restrictions of focus that hamper any photographer working in the 'macro' scale, there is the overriding consideration that must be given to the accuracy of the story. If Barbara's scientific reputation was to be matched by the camera, then it too would have to present a true portrayal of the spider's lifestyle. Happily, I already had absolute confidence in Mantis Film Productions – namely Jim Frazier and Densey Clyne – a team with acknowledged expertise in the specialist field of macro-photography.

Jim and Densey's reputation lies in their attitude to wildlife. The fact that they have developed special close-up techniques and a photographic 'bench' on which they can film the smallest detail of a flower or insect may be a measure of their craftsmanship. But the real magic lies in their enthusiasm. Even the garden of Densey's home is a kind of miniature safari park in which specimens wait their turn for her to bring them before Jim's all-seeing lenses. Inside the house itself you may find Jim working on a 'time-lapse' sequence that will eventually show strawberries growing to ripe maturity in the space of 30 seconds. Then the camera zooms across the berries onto a jar of jam: its label reveals that his little wildlife adventure is a commercial spot for Australian TV. In another corner of the room Jim is working on his reconstruction of the underground home of an ant colony. This is the theatre in which he will shoot the second of our co-productions with ABC: about Australia's bulldog ant, the scourge of many a bush picnic. To gaze into Jim's beautifully structured formicarium with its well-established colony of

BELOW *Bulldog ant-worker offering trophic egg*
BOTTOM *Bulldog ant-worker with larvae*

ants is a special delight. With his camera trained on this underworld, the queen at its heart, he will record how queen and workers extract eggs the size of pinheads from the rear of the abdomen and place them neatly to one side for the worker ants to feed and tend. All this will happen inside an environment of polystyrene built as an exact replica of a real colony. These artificial corridors are fitted with fibreoptic lights sculptured into the roof cavities. The ants will live here for 18 months behind glass while Jim and Densey film them going about their daily tasks in the miniature catacombs.

One thing is clear: we shall not be able to film our spiders quite so easily. The bulldog ant can adapt to the unusual conditions of suburbia with no apparent difficulty. Our spider is not so easily fooled. Although common across a wide belt of desert-like conditions, *Anidiops villosus* has developed a very special requirement for humidity

and temperature which is difficult to match for long periods. Both Jim and Densey are adamant: 'To tell the story of this spider, we're going to have to take the cameras to the spiders even if it means crossing Australia to do it . . .' With the bulldog ants still busying themselves in the background, we get out our diaries to fix filming dates. The sequences we need to film become the centre of discussion:

Young spider emerges from mother's nest
Young spider selects new site for own nest
Starts to dig nest hole
View of digging from below ground
View of digging from close to surface
Mature spider raises lid and captures prey
Mature spider from underground . . . etc.

The shot list was as impressive as it was endless. Barbara, always the wary scientist, was soon convinced that these two macro-photographers really did know their biology and that as well as being experts at their job, would be sensitive about working with such a delicate subject. If mutual confidence between field scientist and photographer is a cornerstone of wildlife film-making, then these three enthusiasts had certainly established a workmanlike foundation.

It was the practical business of filming at Barbara's study site that presented problems. In a landscape where spiders are so easily available it may seem strange that a wildlife photographer should need a controlled environment in which to film. Well, the Australian bush has a lot to offer: except one of the most basic of all essentials – electricity. For six weeks Jim needed 10 or 15 amps to supply light sources and to recharge batteries. Under the circumstances we knew it would be downright foolish to attempt to film on location at night. Apart from the cold chill of the wodjil, the camera lenses easily mist up in rapidly changing temperatures. Then there would be the expense of a portable generator and the frayed tempers as tepid coffee accidentally tips into a freshly loaded film magazine! Of course, we would shoot some sequences when the ABC crew, headed by Don Hanran-Smith, came from Perth to film Barbara on her study site; but the offer made by Val and Ron York to use their garage at the family homestead with its power sockets was too good to miss. For the detailed filming, we would bring spiders to the garage, make a substitute wodjil in a wooden box, then plug in the lights and wait around the clock for something to happen. Well, that was the theory anyway! So it came about that on a Saturday in late June 1980 a heavily laden van bearing on its roof lengths of timber and hardboard with sheets of glass, cameras, lights and two wind-tanned Sydneyites inside, drove out of the sands of the Nullabor plain heading west to Bungalla.

Off-loaded and rested, Jim and Densey settled down into their spacious garage workshop alongside their newly assembled spider environment. About 120 cm (4 ft) long and 90 cm (3 ft) wide, it housed a dozen or so translocated spiders. Some had been carefully reinstated inside their silk-lined living quarters, complete with trapdoor entrance and twig lines. Others had been evicted from their original homes so that the cameras could record them rebuilding their accommodation. Jim supervised the construction of the environmental extravaganza. The sandy earth had to be carefully built around the gossamer-silk nest and firmly bedded down to simulate the compact conditions of the natural habitat. Even leaves of acacia and casurina dressed the surface so that the

Nests of trapdoor spiders exposed

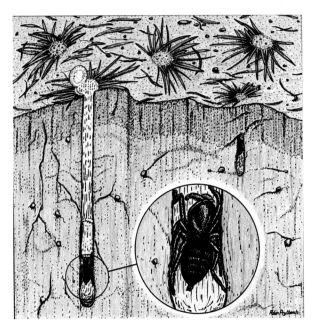

Spider by its trapdoor

setting was indistinguishable from the real thing.
By spraying a fine mist of water to simulate gentle
rain and by controlling ambient temperature and
the brightness of the film lights, they found they
could stimulate the new inhabitants to perform a
diversity of authentic behaviour. Barbara's records
had shown that the favourite time for young trapdoor spiders to emerge from the parental nest is on
cool autumn nights after rain. At such times there
are few predators about: no scorpions or centi-
pedes to devour the 20 or so juveniles that may
survive their first nine months as hatchlings in the
underground nursery.

Our real interest lay in filming the events that
take place during the vital two hours after the
young emerge. Within that short time, in the total
darkness of the wodjil, creatures no longer than a
human thumbnail dig out a shallow depression in
the damp soil and cover it with a hinged trapdoor
entrance. And if that is not enough, each eight-
legged apprentice uses spinnerets at the rear of the
abdomen to bond the fragments of earth together.
Along one side, and one side only, it spins a hinge
of silk to the trapdoor. Most spectacular of all, Jim
was able to film the making of the nest itself, showing how the spider cradles small piles of earth with
its front legs and carries each load up to the surface; each journey getting longer as the nest tube

deepens. For Jim to film the making of the fan-like
veranda at the entrance seemed too much to ask,
but we would have to try. The fan of twigs and
leaves is the sounding board with which each
occupant detects passing prey. As an insect
marches across the immaculate doorstep, it signals
its presence to the spider concealed beneath the
door. The deception is total, passing insects are

66

unaware that they are signing their death warrant. *Anidiops villosus* strikes with poison fangs raised for attack from beneath the forward part of the abdomen. The poison can be serious to particularly sensitive humans; at best it is a painful experience. But for the insect prey, the fangs deliver instant oblivion.

Jim Frazier's photography told this story with vivid authenticity. Not least among his achievements was to show a trapdoor spider actually in the process of attacking its prey. Underground shots showed a mature spider waiting below its door with the insect walking along the twig lines. The film showed that from the moment the spider responds to the movement of the prey, it is a mere one fiftieth of a second before the prey is pulled below the trapdoor for subsequent dispatch – at the spider's leisure! This is not to suggest that *Anidiops villosus* always performed on cue or to order. Often Jim and Desney would wait all night for just one brief shot lasting only a few seconds. One particular resident wilfully refused to spin the lining to its nest whenever Jim's lights were switched on. On another occasion, a breakfast-time inspection of the garage habitat might reveal that a female had constructed the most superbly engineered twig-lined trapdoor in the very hour that Jim, totally exhausted, had drifted off to sleep!

Slowly and painstakingly, Jim and Densey began to piece together the solitary and totally dark existence. The nest burrow itself was, of course, the centre of attraction. Once complete, immature spiders use it as a summer retreat to aestivate – the summer equivalent to winter hibernation – away from the desiccating heat at the surface. As seasons pass, individual spiders repair and enlarge their homes. Jim's film showed the construction of the false bottom of the nest and how, at the approach of an uninvited guest, they instantly pull the drawstring that isolates them at the bottom of the silken tube. It is in this elaborate 'sock' that they store food and unwanted remains of their prey. In turn, this waste will be laboriously carried to the surface and dumped just outside the door. Once the female is mature – probably in the seventh year – she will venture from the nest to find a mate. She will then return for the long fast while incubating eggs and the young brood. This cycle can be repeated every two years from the one mating. Females can be reproductively active for as long as 20 years and

Barbara's observations show that even that ripe old age may not be the limit of their lifespan.

It was the nest construction itself, however, that was the most fascinating. One night, a beautifully marked female chose to perform right in front of the macro lens and Jim captured every detail of the spinning technique. He watched the spider's spinnerets exuding the thin thread of silk and weaving it back and forth across his field of vision. If this expected defeat ended in triumph, it is fair to say that Jim began to despair over the spider's refusal to be filmed building its twig-lined sensing mechanism outside the front door. Even when the time came for our departure, Anidiops had stubbornly refused to display this exceptional skill. We came up with an intriguing, if unorthodox solution. The key to eventual success lay inside two large, black pastic refuse bins that were loaded aboard Densey's van. Each contained a complete nest with resident spider packed tightly with damp soil. Five days later they had travelled 4800 km (3000 miles) east across Australia to become, I suspect, the first live specimens of *Anidiops villosus* to take up residence at Turramurra, just north of Sydney. With recollections of the success of 'Badgerwatch' and the fox-watching exercise fresh in my memory, it seemed just possible that an infra-red sensitive electronic camera, with its very much reduced need of light, might reassure the spider that it was safe to come out, while still being just sufficiently lit to record an adequate picture. It worked. Something that Barbara had seen once only in her laboratory in Perth had been recorded on tape within a week of the cameras being set up at Densey's studio. The pictures were not, it must be agreed, of the most sensational quality, but they were among the most revealing of Anidiops' very special behaviour. There it was, lifting the trapdoor, then pausing as eight eyes briefly scanned the view. Slowly, this gentle and fragile creature gathered up a fine leaf only 50 mm (2 in) long, but for us the equivalent of a large log of wood, then in 10 seconds, trundling its cargo across the surface of the desert, it fondled it precisely into place. If this was not something to wonder about then I had indeed wasted the BBC's air fare to Western Australia. To me, and I hope to the viewers too, this sequence remains a kind of magic, thanks to Jim and Densey, their cameras and the contents of two plastic dustbins!

3. CREATURES OF INLAND AND COASTAL WATERS

THE REAL MR RATTY *producer Maurice Tibbles*

All was a-shake and a-shiver, glints and gleams and sparkles, rustle and swirl, chatter and bubble. The Mole was bewitched, entranced, fascinated. As he sat on the grass and looked across the river, a dark hole in the bank opposite, just above the water's edge, caught his eye, and dreamily he fell to considering what a snug dwelling place it would make. As he gazed, something bright and small seemed to twinkle in the heart of it, vanished, then twinkled once more like a tiny star. Then, as he looked, it winked at him, and so declared itself to be an eye; and a small face began gradually to grow up round it, like a frame round a picture. A little brown face, with whiskers. A grave round face, with the same twinkle in its eye that had first attracted his notice. Small neat ears and thick silky hair. It was the Water Rat!

So begins the encounter between Mole and Ratty in *The Wind in the Willows*, at the start of one of the most enchanting animal stories in English literature. Yet, despite his keen observation of riverbank animals, the author Kenneth Grahame made one mistake: in real life the energetic 'Ratty' is not a rat at all, but one of Britain's smallest aquatic mammals, the water vole.

With the exception of the Isle of Man and much of the Scottish Highlands, water voles live throughout Britain, but they are such timid animals that few people ever see them. The slightest disturbance is enough to send them diving for safety, and all that betrays their presence

The water vole never strays far from water

are ripples spreading out over the water. The water vole's amphibious nature sets it apart from its closest relatives, the smaller bank voles and field voles, which live in woods and damp grassland, and rarely take to water. The water vole depends on the river for safety, and in marked contrast to the fictional Mr Ratty, seldom risks venturing more than 3 or 4 m (10-13 ft) from the safety of the riverbank.

Water voles are among the oldest established mammals in Britain. Fossils show that they have been abundant for about 300,000 years, except during the ice ages when the glaciers drove them south.

The water vole is strictly vegetarian, and like most plant-eating animals, it has to feed almost continuously. Its bright orange incisor teeth cut down prodigious quantities of vegetation, such as grass and water plants, which it then crams into its mouth with its forepaws, while sitting on its haunches, in much the same way as a squirrel deals with nuts. Its body, portly and thickset, is covered in shaggy brown fur which is constantly groomed. Any animal that spends half its life in water must ensure that its fur is in good condition, and the water vole is fastidiously clean, using its tongue, teeth and paws to comb its coat and keep it waterproof. The vole's face is wide and blunt, rather like a hamster's, and its small, brown, bead-like eyes, which can only see clearly at very close range, are set deep in facial fur. Its ears are short, hardly projecting out of its thick coat, but they are very

sensitive, and this, together with an excellent sense of smell, makes up for the vole's very poor eyesight.

Although the water vole is about the same size as a brown rat – around 20 cm (8 in) long – the plump vole and the much more slender rat are easy to tell apart. As well as being different in shape, the rat's tail is long and scaly, whereas the vole's is quite short and is covered with downy fur. Unlike the water vole, the brown rat is a newcomer to the British Isles, arriving from the continent of Europe in the early eighteenth century and spreading rapidly. This aggressive and cunning rodent has proved to be more than a match for the timid water vole. Brown rats are natural scavengers, and can live on a wide variety of food, whereas the voles can only eat plants. When rats invade a riverbank, they often drive the voles away completely and take over their burrows.

Home for the water vole is an elaborate system of tunnels excavated in the soft mud of the riverbank. As in all its movements, the vole's digging is rapid, and it scrapes with its feet while pushing surplus earth out with its head. The banks that water voles colonize are often riddled with holes, and it is easy to understand how they acquired their old country name of 'water mole'. Each burrow usually has at least three entrances, one of which will be below

water level. This is used as an emergency entrance, and to conceal its location from predators, the vole has evolved a remarkable technique for covering its tracks. Just before it swims into the submerged hole it stirs up the water with its forepaws, creating a muddy smokescreen which hides it from view – the perfect vanishing trick.

The water vole swims both on the surface and underwater. Surprisingly for such an aquatic

RIGHT *and* BELOW *The water vole swims both on the surface and under water*

animal, its toes are not webbed, but its rapid paddling propels it forward quite swiftly. If the vole is alarmed, it will plunge into the water with a splash to alert other voles, but normally it enters the water with little more than a gentle 'plop'.

Restricted by its timidity, the vole prefers to feed close to home, nibbling all the vegetation within reach of its burrow. This creates a water vole 'lawn', a distinctive area of closely cropped grass near the feeding hole. When this has been trimmed to the ground, the vole is forced to look further afield. Often it will simply travel along the bank, creating well-trodden waterside paths that lead to the lushest vegetation. Leaves, stems, roots and corms are all eaten avidly, and lily stalks are one of the vole's favourite foods, but what tempts a water vole more than anything else is an apple. The smell of an apple core floating downstream quickly brings out the vole who seizes the prize and carries it back to the bank to be eaten.

The vole's lack of adventurousness leads it to use the same paths again and again, and in water it sticks to the same channels between weeds on its regular forays. On land the vole marks its runs with a scent produced by glands on its flanks. This not only helps to guide the short-sighted animal back to its burrow, but also indicates to other voles that they are intruding on an already-claimed territory.

In early spring, as the riverbank vegetation begins to sprout once more, male voles that have managed to survive through the winter begin the search for a mate. Each has a small territory surrounding his burrow, and an increasing amount of time is spent inspecting its perimeter. The borders of his riverside domain are marked by latrines, where little piles of droppings signal that he is in residence. He also marks the latrines with his scent glands to warn off any other males that might be on the move. Females, however, are tolerated. When one enters the male's territory, he courts her in a brief and almost casual way, and mating takes place either on land or in water. After mating the female generally moves away, and the male searches for further mates. The female has four weeks in which to find an unoccupied stretch of riverbank, dig a burrow and then line it in preparation for the birth of her young. It is a busy time as she feeds and collects sedges, fibrous roots and the pithy core of rushes to insulate the underground chamber that will house her brood. By the time she is ready to give birth in late April the riverside growth is at its most luxuriant.

When the vole's young – usually about five in number – have been born, she leaves the nest. At this time of the year she eats only fresh food, storing nothing, and she must therefore leave the nest to feed if she is to provide enough milk for her young. Her frenzied feeding is interrupted about four times a day for periods of suckling and cleaning her 50 mm (2 in) long offspring. The young voles are naked and blind, and are entirely dependent on their mother.

A sudden storm can bring disaster because the nests are often near water level: a rise of just a few inches can flood the burrow and drown the young. As the water level rises, the mother vole scampers back to her nest, and if the burrow is threatened, she will attempt to save her family by carrying them one by one to a higher chamber. If she fails, she will mate again and raise another family.

The life expectancy of a young water vole is just one brief summer – on average they will live no more than five and a half months, although some veterans may survive for a year and a half. It is a short time in which to develop, mature and reproduce, but voles do not take long to reach adulthood. A week after birth they are covered in fine red hair, and they have doubled their weight. Their eyes do not open until the eighth day, but only about a week later they leave the safety of the nest for the first time to feed in the dangerous world outside. Initially their emergence from the burrow is brief and hesitant, but they quickly grow in size and confidence, and in just one month they are ready to fend for themselves. As they are weaned and develop the darker fur of mature animals, the young take over the territory, brusquely driving away their unfortunate mother. Some females mate again and give birth to another litter, but often they perish in the autumn, weakened by weeks of intense activity raising their young.

Despite their short lifespan, with this very rapid rate of reproduction, water voles could theoretically overrun all the rivers, streams and drainage ditches in Britain, but the activities of predators prevent this from happening. The silvery glint of a passing vole quickly attracts the attention of any pike that may be lurking close to their burrows and this freshwater predator has no difficulty catching its prey. With a single flick of its powerful tail, the

Heron on the look out for water voles

pike approaches the vole, and a swift snap of its jaws effects the kill. The shallow waterways favoured by the voles are also ideal for herons, and many voles meet their end in the sharp bills of these fishing birds. Voles also have much to fear from carnivorous mammals. Foxes and stoats regularly take their share; and the mink, a ferocious predator introduced from America and now quite common along English rivers, can catch the voles both on the ground and in the water.

Voles are especially vulnerable because they can hardly see underwater at all, and they have no real defence against predators except to swim for the security of their burrows. Even there they are not always safe, since stoats can pursue them underground. The vole's keen sense of smell can warn it of the whereabouts of most of its enemies, but even if it is aware of a predator's presence and retreats to its burrow in time, the vole cannot stay underground for long, or it risks starvation. And there is one predator whose attack is so sudden and unexpected that the vole has no time to detect its approach: the barn owl, whose soft plumage muffles the sound of its flight. It only needs the slightest sound or movement to locate its prey, and the voles are in danger whether they feed by day or night. From a vantage point above the river, the barn owl swoops with an easy glide and silently drops on to its victim. Faced with all these predators, the water vole needs its great reproductive capacity simply to survive.

For those that escape the daily danger of being eaten, the first frosts signal the beginning of a new struggle. This time the adversaries are cold weather and a lack of food and few voles will win this contest. The thick vegetation of the riverbank dies back, and the voles are forced to travel further and further for their food. Far from retreating to the warmth and comfort of their burrows like Kenneth Grahame's Mr Ratty, they must be up and about, running the gauntlet of their hungry predators. For most surviving parent voles, it means that their brief lives are drawing to a close, and the same is true for those young born late in the summer, since they have not had sufficient time to build up their body weight in preparation for the winter months. For Mr Ratty's relatives in the real world, winter is the ultimate predator.

L.G.

OPPOSITE *Voles make ideal prey for the barn owl*

BELOW *The smell of an apple attracts the water vole*

GENTLY SMILING JAWS producer Adrian Warren

How doth the little crocodile
Improve his shining tail
And pour the waters of the Nile
On every golden scale.

How cheerfully he seems to grin
How neatly spreads his claws
And welcomes little fishes in
With gently smiling jaws!

from *Alice's Adventures in Wonderland*
Lewis Carroll

'Sinister', 'repulsive', 'unpredictable' are words often used to describe crocodiles. Certainly they are ancient creatures that have changed little since their ancestors lurked in the quiet waters of tropical lakes and rivers in the age of the dinosaurs, some one hundred and forty million years ago. Although there are crocodiles in America, Asia and Australia as well as in Africa, none has achieved such fame or notoriety as the Nile crocodile. The accounts of early explorers were filled with superstitious awe and dread for these reptiles,

with their cold, indifferent eyes, that can grow up to 5 m (17 ft) long and weigh close on a ton. But the Nile itself, the longest river in Africa, no longer teems with them; in all but a few places they have been hunted out of existence either because they are considered dangerous or, more often, for their hides. In other parts of Africa they are more numerous, for the Nile crocodile is not only limited to the river from which it takes its name, though everywhere it is rapidly declining in numbers. This is the reptile once worshipped and mummified for posterity by the ancient Egyptians, still venerated in some monasteries in Pakistan, and whose reputation as a killer rivals sharks, forcing people to fear it in the extreme.

Until only recently our knowledge of the Nile crocodile was almost wholly based on garbled and exaggerated legends. The many accounts of attacks on human beings have resulted in almost universal revulsion towards this poorly understood animal. The fact that the crocodile has survived the ages at all is by virtue of its perfect adaptation to its river and lake habitat. Until a decade or so ago, however, this success was very difficult to explain on the

LEFT *Nile crocodiles*

OPPOSITE ABOVE *The crocodiles leave the water to bask in the heat of the sun*

OPPOSITE BELOW
Tony Pooley

basis of the animals' recorded natural history. The consensus of observations made by hunters, missionaries and explorers was that the Nile crocodile was a lethargic hulk that spent most of its life basking in the sun, only now and again rousing itself to attack an unwary animal at the water's edge.

Tony Pooley is a South African naturalist who has made a study of wild crocodiles over the last 20 years. His interest in them began in 1958 when he was working as a game ranger in Mkuze Game Reserve in Zululand, a district of Natal. Accompanied by some game guards he found a crocodile's nest near the Mkuze River. As he had been unable to find satisfactory answers to questions such as what is the incubation period, size of the eggs and size of the young crocodiles at hatching, he collected the eggs and brought them back to camp where they successfully hatched two months later. Suitable ponds and enclosures were built for the hatchlings and thus began a study that has absorbed him ever since. Crocodiles were observed

ABOVE *Female with young in mouth showing her pouch*

OPPOSITE *The young mill around her in the water*

in the mangroves and estuaries, in open water, in swamps, in fast-flowing rivers and in the quiet lagoons. Nesting grounds were mapped, and more and more Tony's interest began to focus on their breeding biology; this was an area on which few writers seemed able to agree and there were even suggestions that crocodiles eat their own young.

Crocodile nests are holes in the ground, dug by the mother, perhaps 30 cm (12 in) deep, occasionally some considerable distance from the water. One of the mysteries Tony wanted to solve was how the youngsters escaped from the nest and reached the water. He had observed when collecting eggs that the clay soil can bake so hard that he had to chip out each egg with a knife; the young therefore could never penetrate this thick, hard layer and reach the surface unaided. Parental assistance, Tony argued, was essential, and to back up this theory he had noticed that when the young were ready to hatch, their yelps and croaks as he approached the nest were audible from some distance away.

An early observation in 1774 by Oliver Goldsmith was that 'On being set free the brood quickly avail themselves of their liberty, a part run unguided to the water, another part ascend the back of the female, and are carried thither in greater safety.' To try to solve some of the mysteries of crocodile parental care, Tony started to collect adult crocodiles, housing them in huge pens at Ndumu, a game reserve in the far north of Natal. By October 1973 he had nine adults in residence including a female that was guarding her nest not four metres from the fence. This was the opportunity Tony had been waiting for and he visited her three or four times a day. During the first month Tony was greeted by a baleful glare and low throaty growl, but as time passed the female became accustomed to the visits and by the end of the second month would not even turn her head if a noisy group of strangers came near the fence. She guarded the nest closely: in fact, during the first 117 days of observation, she was only seen to be absent from guard duties twice – both on particularly hot days – and she soon returned to the nest after Tony arrived. But past nesting studies had indicated that the incubation period for the eggs was, at the most, 90 days, and before long the

female began to lose interest in her nest. It was obvious that the eggs were either rotten or infertile.

Tony takes up the story:

'I decided to test her reactions and took along a box of live young and unhatched eggs, as well as a tape recording of the sounds that the young make when they hatch. I had no idea what would happen but to my surprise as soon as I switched on the tape recorder and played the calls, the reaction of the mother was instantaneous. She came straight to the fence, trying first to dig her way under it then to push her snout through the wire, obviously determined to trace the source of the sounds. The next step was to offer her the live youngsters to see what she would do; I was a little horrified when I did this because the first one I put through the fence she seized between her teeth and literally gulped it into her mouth. I thought, as some people had suggested, that maybe adults do eat their young. After watching her for several minutes, I heard a muffled croak from inside her mouth indicating at least that the youngster had not been swallowed. Then she lifted her head to see if I had more; it was then that I noticed that beneath her jaws a huge pouch was hanging down that I had never noticed

before on any crocodile. I decided to take the plunge and quickly offered her more youngsters through the wire; in turn, these were picked up one by one. In fact some of them actually walked towards the head of the female and advertised their positions by standing up almost vertically with their front legs outstretched, tails swishing from side to side and croaking loudly to attract her attention. And when she bent down to pick them up, some of them voluntarily climbed over her teeth and flopped into the pouch. Then I gave her eggs, unhatched eggs, and these too she picked up and gulped into her pouch, and the whole lot were carried down to the water. And when they got to the water, the young were reluctant to leave the security of the female's jaws, in fact she had to wash them out by vigorously swishing her head from side to side to force them out. Once in the water the youngsters milled around her, climbed on to her head and her back, obviously not wanting to leave her. She had demonstrated to me that memorable day that parental care in the crocodile is very strong; it was now fact, not fallacy.'

Tony's discovery was an important one for it showed that the massive jaws and murderous teeth capable of crushing prey with a single snap can, like a pair of forceps, delicately pick up the youngsters and gently carry them to the water. And this was just the beginning of a series of discoveries made by Tony, revealing the Nile crocodile to be a complex, intelligent animal, quite the opposite of its traditional image.

The story of Tony Pooley and his crocodiles had all the ingredients for a natural history film, but to shoot it was a daunting task which fell into Rodney and Moira Borland's laps. Rodney and Moira are a South African-based, husband-and-wife team who have been filming wildlife for over ten years. They had agreed to shoot some programmes for 'Wildlife on One', and in fact had just completed their first one on the white rhinos of Umfolozi. The problem with the crocodile story was that it would have to be shot over a year to record the full cycle of behaviour from courtship through egg-laying to hatching and finally parental care; and we also hoped to accompany Tony on a crocodile hunt and to reconstruct his remarkable experience with the mother croc by the fence at Ndumu.

The first day Rodney and Moira visited Tony at his research centre, crocodiles to them were still sinister animals to be given where possible a wide berth. But within a few hours, Tony was able to captivate them with these remarkable animals. He took Rodney and Moira to the huge enclosure where he keeps 18 large crocs all captured because they were trouble-makers, either taking livestock or threatening human life. As Tony explained, a crocodile is a large predator that does not differentiate between animals and humans; anything of the right size that wanders too close to the water's edge is potential food. Some of Tony's crocs were in the water, others were on the bank, jaws agape, sunning themselves. Tony explained that they are all well fed, but still need to be treated with respect: each has its own threshold distance beyond which it would be dangerous to pass. At that, Tony hopped over the fence into the enclosure and invited Rodney and Moira to join him. 'After a while they get to know you and accept you,' Tony explained. 'As long as I am here, you'll be quite safe.' And to prove it, he sat on the ground and called to his crocodiles. One by one they came, some from the far side of the enclosure, a distance of perhaps a hundred metres or more, swimming across the pool and assembling themselves on the

ground near Tony. For Rodney and Moira it was an eye-opener; these crocodiles not only accepted Tony but they actually seemed to like to be near him. After that Rodney and Moira were sold on crocodiles and determined to make a film that would be a good public relations exercise for these now-not-so-sinister reptiles.

With characteristic enthusiasm, they started to watch crocodiles from dawn until dusk just as Tony had done in the early days. And they soon found that hides, normally an essential piece of filming equipment, were useless. To study crocodiles you have to creep through the reed beds and watch from a quiet vantage point. Crocodiles are suspicious of hides and keep away; the only way to

TOP *Crocodiles cool themselves by gaping*

ABOVE *Too hot they move into the water*

OPPOSITE *Rodney and Moira Borland*

80

watch them is to make yourself obvious by sitting in full view of them. They can then see you and after a while emerge from the water to bask, and after a few days they totally accept you. Quite a large percentage of a crocodile's life is spent doing very little. They leave the water to bask and absorb heat from the sun. Like all reptiles, they are dependent on the environment for their body temperature, although they can to a certain extent control it through their behaviour. So as soon as the sun rises, they come out onto the riverbanks to bask and several hours of the day will be spent in this apparently torpid state, although the slightest disturbance brings them back to life. There are no pores in the thick armour-plated skin so crocodiles

cool themselves by gaping, allowing cool air to waft over the soft skin inside the mouth. When they become too hot they move into the shade or back into the water to cool off.

Tony needed another adult male crocodile for his research programme and suggested that Rodney and Moira might like to film the hunt. The best time to hunt crocodiles is at night, preferably a pitch-black, moonless one so the crocodiles cannot see the outline of the boat. So Rodney and Moira started working nights too; after a long day watching and filming crocodiles basking and feeding they would meet Tony by the jetty, pile the equipment into a tiny boat and set off into the blackness. They were probably wondering what on earth they

had let themselves in for; after all, how do you persuade an angry, writhing croc, jaws and all, to share a tiny boat full of people and equipment in the middle of the night? For several nights they caught nothing. For hours on end they would punt quietly along the river using a powerful torch to find the crocodiles, whose eyes gleam like red coals in its beam. Sneaking slowly towards the eyeshine, Tony would try to slip a noose over the head before it submerged. A few times he almost did it, but the crocodiles kept sliding skilfully out of the noose before Tony could draw it tight. But finally he managed it, and the moment the noose was pulled tight and the crocodile trapped was the most dangerous. An enraged crocodile weighing some hundreds of kilograms could easily overturn a boat. So it had to be played carefully but Tony's experience and skill made the operation look easy. First, the jaws were secured with ropes, and then the tail. the croc was hauled into the boat and the journey home began.

At the pens the following morning, Tony released the crocodile from the ropes. It was unlikely to have much energy to attack since it had been out of the warm river water all night and was therefore cold. As a croc's body temperature falls so does its ability to move quickly. The newly captured croc slowly slithered off through the grass into the pool.

Tony Pooley's research programme is now not so much geared towards studying crocodile biology but to rearing youngsters for release into the wild. As Tony says, too many wild crocodiles are being killed to become handbags or shoes. Yet crocodiles play an important part in the ecological scheme of things. They are as important a predator in the waterways as a lion is on the plains keeping down antelope populations. In the lakes and rivers, large fish like barbel eat almost everything else. They eat molluscs, frogs and their tadpoles and the roe of certain kinds of edible fish; they eat crabs and the fingerlings and adults of other fish, and they breed prolifically. Their only predator is the crocodile. If you eliminate the crocodile from the natural system, the barbels take over, virtually exterminate all other species, which in turn affects the growth of plant life, and many species of aquatic birds which can no longer find the food they need. In other words the whole ecosystem is disrupted. So Tony started releasing crocodiles back into the wild, in areas where he believed their future to be

threatened. Some of those he released would, he knew, ultimately be hunted; but some, he hoped, would survive.

The success of introducing the newly captured male into the pen depended on how quickly he was accepted into the social hierarchy of the other 18 crocodiles in the enclosure. But within a few weeks he was feeding normally and eventually selected a mate. Perhaps surprising for these ancient creatures, courtship is a complicated affair. It is a process always acted out in the water and it is often the female that lures the male down from his basking spot to mate. The male, at first, approaches her slowly, circling her, gently bumping her, and then rubbing the underside of his jaws over the head of the female. It is not certain but it may be that the male is produces a perfume from glands under his chin which stimulates the female to the next stage of courtship. Mating seems extremely laborious, even though the water gives their bodies buoyancy; the male lying on top of the female and both twisting the rear ends of their bodies together. It is a remarkably passionate display for supposedly sinister animals, and one wonders if the crocodile's gigantic ancestral relatives, the dinosaurs, were perhaps just as gentle in their passion.

Two months later, the female starts looking for a site to lay her eggs. As Tony had observed, crocodiles have communal, ancestral breeding grounds and normally return to the same spot year after year to lay. These traditional nest sites are chosen because they have best chance for success; but even so there is still a surprisingly high failure rate. Some 60 per cent of nests in Zululand are destroyed by predation or through flooding and only a tiny proportion of successful hatchlings will survive to parenthood.

The female begins to dig her nest hole. Using her powerful hind feet she scoops out the soil to a depth of about 30 cm (12 in). It is a long process and the hole has to be just right. Once she is satisfied, she deposits about 50 oblong leathery white eggs, each about the size of a chicken's egg. Still in the laying position, she pushes the soil back into the hole and neatly packs it down firmly with her hind feet. From now until the eggs hatch she will guard the nest day and night, not even leaving to feed, for just over three months.

For Rodney and Moira, the filming was going

Crocodiles cannot chew but gulp down large fish. With larger prey the teeth are used to rip off chunks small enough to swallow

Crocodile courtship takes place only in the water

RIGHT *Newly hatched crocodile*

OPPOSITE *The birth of a crocodile*

well but the hatching and parental care was of course the climax, so as the time drew near, their excitement rose. **Rodney and Moira describe their daily routine at that time:**

'When the due hatching date arrived we started our vigil. The wretched alarm would go off at 4.30 am with its ghastly electronic *zi-ing* in our ears and one of us leapt out of bed, in case we fell asleep again, and put the kettle on. A quick cuppa and we were off to arrive at the nest just before dawn. We unloaded all the equipment we needed: cameras, lenses, film and tape recorder, in all about six heavy boxes. Then we sat and waited. The days were long, hot and humid, but it was essential to film the mother as she began to dig, so we sat with eyes glued and camera ready. At dusk, when it was too dark to film we trundled home, to eat, clean up and prepare for the following day.'

The days went by and for twelve days nothing happened. Then, quite suddenly, they got it. There were muffled sounds, croaks, from below the ground. The ever-patient mother became alert, approached, listened again, then came right up to the nest and started scraping away the hard baked earth. Underground, many of the baby crocodiles had already ruptured their egg shells with special egg teeth; when the mother finally dug down to uncover them, they were already struggling free. Then, just as Tony had said, she gently took them between her jaws and manoeuvred them into her

pouch, now bulging prominently beneath her lower jaw. If the young have difficulty in breaking out of the egg, she helped by picking up the egg and rolled it between her teeth with just enough pressure to rupture it without harming the youngster inside. Gradually she transferred them all to the nursery pool she had selected. Although she had not eaten now for three and a half months, she remained with the young crocs for the first few weeks of their lives. If danger threatened, they flock near to her, but little by little they learnt to wander further afield in search of food. Only then did the devoted mother break her remarkable fast to go off and find food for herself.

The croc film was in the can and for Rodney and Moira it was all over; well, not quite – they were making another film for 'Wildlife on One' at the same time as filming crocs, on the ecology of a place called St Lucia. The trouble was it was turning into a very difficult project and the crocodiles had efficiently mopped up most of their time. For Tony it was not all over; he continues to study crocodiles and still learns something new about them every day. For Rodney and Moira it had been an extraordinary experience, with surprisingly few nasty moments. Only on one occasion in the entire year did a crocodile fancy eating Rodney and he had been able to back away in time. The worst experience occurred while filming crocs underwater, not with a croc but a large leech that swam into Rodney's ear and had to be removed with a pair of pliers.

ST LUCIA - LAKE WILDERNESS *producer Adrian Warren*

To most people, St Lucia is a tropical island in the Caribbean, but to someone from South Africa, St Lucia is a very different place. The African St Lucia is a vast complex of waterways, reeds and mudflats in Zululand on the edge of the Indian Ocean; an aquatic wilderness the size of Greater London. It is a spectacular wildlife refuge and although the southern end and St Lucia village itself are well-known as a fishing and holiday resort, large parts of the lake are kept undisturbed with minimal human contact. Here the variety of birdlife is particularly rich and was even recognized as such almost one hundred years ago when St Lucia was declared a game reserve in 1897. It was the first game reserve in the whole of Africa.

Imagine our disappointment then when, having

Hippo at sunrise, St Lucia

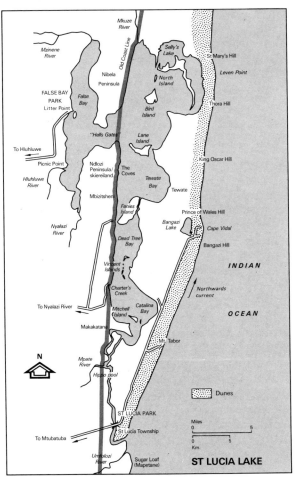

decided to make a programme about St Lucia, we
found great difficulty not only in filming the wildlife but in finding anything to film at all. Not that
the wildlife was not there, but it was a case of the
cameraman having to be in the right part of the lake
at the right time of day or season, and this sort of
knowledge only comes slowly. The very vastness
of St Lucia did not help. It would take cameraman
Rodney Borland and his wife Moira several hours
to drive from one end of the lake to the other in
their Landrover, over bumpy, tiring dirt roads; and
to find nothing worthwhile to film once they had
arrived become demoralizing after a few weeks.

The filming of crocodiles for 'Gently Smiling
Jaws' had to go on and was making heavy demands
on their time; it also obliged them to make their
base at the southern end of the lake in case they had
to be called out in the middle of the night to film a
crocodile egg-laying.

It soon became clear that while the crocodile film
was making excellent progress under the guidance
of Tony Pooley, the St Lucia story was lagging
behind. Any natural history film about a place is
difficult; much depends on luck and patience, and
it is therefore almost impossible to script. All the
producer can do is offer guidelines to the cameraman in order to try to keep within a storyline; these
guidelines are based upon other people's observations and occasionally, if one is lucky, on a short
visit to the location to discuss shooting plans with
the cameraman.

St Lucia undoubtedly had a fascinating story to tell but it was a hard nut to crack. Its story began a hundred million years ago when the play of ocean currents and winds along the low coastline of Zululand deposited sand and silt that eventually grew into a chain of sand dunes rising in a long line of hills just behind the seashore. The process fed on itself, and as more and more sand piled up, some dunes rose to heights of 120 to 180 m (400 to 600 ft), protecting this vulnerable area of coast from erosion by the sea. Dunes furthest inland became covered by pioneer plants able to tolerate drought and frequent drenching by salt spray, and their root systems bound the sand fast preventing it from being blown away. Along one stretch of coastline about 150 km (93 miles) south of the border with Mozambique, four rivers once ran directly into the sea, but their flow was effectively staunched by the massive dunes. Behind this impenetrable barrier therefore a huge shallow lake formed, fed by the rivers, and connected to the Indian Ocean by a narrow estuary at the southern end through which tidal water surges and ebbs.

The turbid, brackish waters of St Lucia were born and became a magnet to wading birds, hippopotamus and crocodiles. The hippos in the lake now comprise the most southerly population in

Africa. There are several hundred of them dispersed in herds of 50 or so, and they play a key role in the lake's ecology. For one thing they defecate in the lake, fertilizing its waters with digested plant matter which would otherwise take months to be broken down and washed in by streams. Wherever there are hippos, therefore, the water becomes rich with small organisms that form the base of a food chain nourishing vast numbers of fish and birds. The hippos' feeding habits have another unexpected value: in eating grass at the lake's edge, it becomes closely cropped making it ideal for ground-nesting birds such as spoonbills and white pelicans, and even the hippos' footprints in the mud provide home for small aquatic animals as they fill with rainwater.

The hippos rarely emerge from the lake except during the hours of darkness to graze; during the day they remain in the water where they feel secure, keep cool and where their enormous bulk derives benefit from the water's buoyancy, relieving them of the effort of standing up. Although on average an adult male hippo can weigh as much as 3 tons, in water they are almost weightless and can move easily by tip-toeing across the muddy bottom of the lake. During the day short periods of argument, play or passion are interspersed with

OPPOSITE *Hippo are rarely seen out of the water*

RIGHT *Saddlebill stork*

BELOW *African spoonbill*

BOTTOM *Goliath heron nearly 1.5 m (5 ft) tall*

seemingly endless periods of rest and inactivity when the hippos are only visible as uninteresting dark blobs on the surface: another test of patience for Rodney as he shifts uncomfortably on his perch in a tree a hundred metres away. He had struggled through the darkness to set himself up there with his camera to film the hippos as they shuffled back into the water, in the pink light before dawn, after a strenuous night's feed. Now if he wants to film them leaving the water again at dusk he must stay where he is all day, hardly moving a muscle so as

not to betray his position. On this occasion Rodney's patience was rewarded; at dusk they did leave the water to feed while it was still reasonably light, and he shot some remarkable film. First a mother emerged with her youngster, then eventually the whole herd. They do not simply blunder ashore anywhere but instead follow old tracks that lead them to their favourite grazing areas. The hippos tear up the grass not with their teeth but with their lips, which are wide and tough, then the teeth are used in chewing. A single hippo may consume up to 77 kg (170 lb) in one night – a whole herd therefore can strip the lakeside of huge amounts of vegetation.

As the hippos stroll back to the water at dawn, St Lucia's fishing birds preen themselves in the first warming rays of the sun before beginning the day's fishing. Because few parts of the lake are of any depth, herons, storks and egrets are able to fish over nearly all of its area. Hunting by stealth they usually feed alone or in pairs. The majestic grey and chestnut-coloured goliath heron, nearly 1.5 m (5 ft) tall, stands as if carved out of stone when fishing, its motionless figure waiting for a fish large enough to be worthwhile to swim within range of its swift, stabbing beak.

The strikingly beautiful and, at St Lucia rare,

saddlebill storks use a different technique. Instead of watching and waiting they steadily wade through the water to disturb small fish with their feet, snapping them up with their bills as the fish swim away. Their beautiful red-and-black bills capped with a broad yellow saddle make them among the most handsome water birds in Africa. They may catch more fish than a goliath heron in a day, but they are likely to be smaller and they have to work harder for them.

In contrast with the stealthy herons and storks, white pelicans are boisterous, social feeders. Each adult needs about 1.8 kg (4 lb) of fish a day, much of which is caught in cooperative fish drives, when a large number of birds join together to force a shoal of fish towards the shore. They gather in a noisy crescent and then simultaneously start swimming landwards, beating their wings and legs to stir up the water as much as possible. If all goes well the fish become trapped and the pelicans scoop them up in their beak pouches, which become grossly distended as they fill up with water; once this has drained off, the pelican swallows its catch.

The white pelicans stay together as a single flock, ranging widely over the lake and roosting on land near the water's edge, often near a hippo pasture. For months, Rodney was unable to obtain a single shot of these birds, though occasionally he would catch a glimpse of them in the distance as they spiralled high in a thermal, the sun glinting on their brilliant white feathers as they turned. The rangers would contact Rodney if they saw the flock, and he would charge off in the Landrover to a distant corner of the lake only to find they had already moved on. But finally on one magical evening when Rodney was filming hippos, they

ABOVE *Pink-backed pelicans*

LEFT *White pelicans*

OPPOSITE BELOW *Reed cormorant, one of the many birds that co-habit with the pink-backed pelican*

90

came to him, a long line of flapping birds almost skimming the water, and as the sun went down they landed one after another not 50 metres away from his camera. After his months of purgatory, Rodney felt he deserved some luck. But in a way, the white pelicans were out of luck too. The bad weather and high salinity in the lake caused their breeding efforts to be a disaster. The pelicans' wandering lifestyle is normally interrupted for the breeding season when they congregate on some remote islands in the middle of the lake to nest, raising their young in shallow scrapes of weed and feathers on the ground. The year of the filming they had tried but nearly all the chicks died.

The high salinity in the lake played havoc with the fish population and many other birds suffered or moved out as a result. The pink-backed pelicans were luckier; they breed at a slightly different time of the year and therefore did not have to raise their young at the critical period. St Lucia is the only place in South Africa where the pink-backed pelican breeds, and the nesting site is closely guarded and protected. Unlike the white pelican, the pink-backed pelican builds its nest in mimosa trees near the water. The nest is an untidy platform of large sticks built in close proximity to others, and indeed the mimosa trees are shared by numerous other nesting birds: herons, egrets, ibis, cormorants and darters.

The pied kingfisher

Each fishing bird specializes in a particular method of catching its prey. The pied kingfisher uses the air as a platform where it hovers, head motionless and eyes transfixed its body and wings working furiously to hold its position in space perfectly. If it spies a fish near the surface, its drops out of the sky like a dart, wings folded for added speed, to pluck the surprised fish out of the water.

The kingfisher's attack can only be appreciated in slow motion, but the high-speed dive was not the easiest thing to film, even when the action was within range of the camera. Having found a good location, Rodney caught some fish and put them in an aquarium. This he took by boat to the kingfisher location and sank it in the water, fixing it so that the rim of the tank was just below the surface; from Rodney's hide, the tank itself was not visible, but from the air, the kingfishers would see the fish and therefore dive predictably to make filming easier. At least that was the plan. What Rodney had not

LEFT *Lesser flamingos*

OPPOSITE *Papyrus*

fully appreciated was the speed of the tide; every five minutes or so the rim of the tank became uncovered and therefore had to be sunk further into the water, and not too far or the fish would escape. Every visit to adjust the tank disturbed the birds and they took a while to settle down again after Rodney had returned to the hide. To crown it all, the daily ranger's boat was heard approaching on the dot of eleven; with a friendly wave to Rodney, he swept passed within a few feet of the tank and the wash uprooted it, leaving it exposed on its side, empty of fish. The film sequence was finally filmed *au naturel*, again several days of patient work for only a 30-second sequence.

The magnificent fish eagle is another bird that dives on its prey from the air, but is more spectacular in its stoop to grab large fish out of the water with outstretched talons. Having worked out the daily routine of a particular fish eagle, Rodney was able to predict where it would be at a certain time of day and another sequence was in the can.

Rodney also took to the air, to film from a helicopter being used by the Natal Parks Board to survey reedbuck in the area. From 1200 m (4000 ft) up it was possible to appreciate the lake's geography; the long line of massive dunes along the coast, the thin neck of water that forms the estuary

at the southern end, the complex network of channels and mudbanks near the estuary and to the north, the vast sheet of water that forms the main body of the lake, some 40 km (25 miles) long and 15 km (9 miles) wide. By skimming low across the lake's surface, Rodney was able to capture a breathtaking scene: sky reflected on glassy smooth water and pastel pink flamingos scattering fan-like away from the helicopter's flightpath. Viewing the lake from the air gave it a new dimension; the collections of blobs that were hippo herds could be seen in relation to one another and the large numbers of floating objects looking like logs were in fact crocodiles.

In protected areas, crocodiles grow to immense sizes and have been known to attack men careless enough to loiter by the water's edge. Their distribution depends on the salinity: if it is high, the crocodiles make for the rivers feeding the lake, preferring the freshwater. But in the breeding season, females congregate at ancestral nesting sites, on ground of the right consistency and high enough to escape any danger of flooding. Strangely enough this super-predator has enemies much smaller in stature. Both water monitor and water mongoose will, if they dare, raid crocodile nests for their succulent eggs, running for cover in the reeds

weaving is complete the males flutter beneath their newly constructed homes to attract a female, who will line the nest to her satisfaction and then lay her eggs.

A third kind of weaver, the finch-like thick-billed weaver, is less ostentatious. The drab brown male and female share the work of building the nest after they have paired, creating a beautifully neat pouch-like structure slung between two reed stems, with an entrance hole just large enough for the parent birds to squeeze through. They tend to nest in areas favoured by crocodiles, which afford them obvious protection from certain predators.

Far out on the lake a party of flamingos sift through the water with their angular bills to collect small animals and plants. Pelicans fly low over the surface, their necks tucked in and their broad wings beating powerfully. A jellyfish swims with a pulsating movement up at the northern end of the lake indicating a high salinity. With such shallow, tidal water, the ecology of St Lucia is constantly changing. In time of drought inland the supply of freshwater dries up, water from the sea flows in and evaporates resulting in very high salt levels. Freshwater animals that can do so retreat, but plants die to be replaced by salt-tolerant sea grass. But those that die add more nutrients to the lake and in a while it rains inland, freshwater pours in and flushes out the salt, allowing yet a new set of plants and animals to colonize the waters of St Lucia. This cycle of events is important to maintain the food-rich waters of the lake; the only danger is that the water could, being already so uniformly shallow, silt up completely. The narrow estuary has been known to become blocked with sand and mud but now this channel is regularly dredged to keep it open, a sort of lifeline for St Lucia, allowing the waters of the ocean and the lake to mingle with one another. As long as they continued to do so, the animals and plants will flourish, and migrant birds will continue to stop over on their long journeys to and from such far-away places as Siberia. A hundred years ago, when St Lucia was declared a game reserve, nobody worried very much about the future of wildlife but they recognized that this small corner of Africa was a remarkable place, as indeed it still is today. For Rodney and Moira Borland, St Lucia will be remembered in terms of several months' painstaking, frustrating work; but always the sense of reward is worth all the toil.

when disturbed by the ever-watchful mother crocodile.

Some of the reedbeds are extensive and some consist solely of papyrus, a giant reed with a feathery mop on the end that grows to twice the height of a man. Papyrus grows so densely that it is difficult to move through and the protection that it offers, combined with its value as a building material, attracts thousands of weaver birds to nest near the lake's shore. When the weavers are nesting, the papyrus comes alive with their movements as they flit between the feathery papyrus heads like large yellow butterflies, calling constantly to one another. Two different species, the masked and the spotted backed weaver, often nest together. They are sociable birds and prefer to nest as close to each other as possible. The male does the initial construction work, trimming off the brush-like tufts of adjacent papyrus heads and tying them together to make a stable platform from which the rest of the nest will hang. At first the nest is just a shapeless knot of grass and strips of reed, but gradually it fills out until the domed exterior is complete. With so many birds building nests at the same time there is much noise and squabbling over materials, and occasionally one will steal a choice piece of reed from his neighbour. Once the

RED RIVER SAFARI   producer Richard Brock

The River Tana begins its 480 km (300-mile) journey to the Indian Ocean in some of the most incongruous scenery in Africa – the snow-covered peaks of Mount Kenya, which lie almost exactly over the equator. Most of Mount Kenya's 5200 m (17,000 ft) height is made up of gentle slopes that fan out for miles around the mountain's base, but at the summit, moorland gives way to steep crags of bare rock as cold and windswept as any peak in the Alps. Mount Kenya was once a huge volcano, but today it is not molten lava that tumbles down its slopes, but icy water destined to feed one of the few rivers in Kenya that flow throughout the year.

The bleak moorlands through which the Tana's tributaries flow are the home of some very unusual plants. The giant groundsel, a 6 m (20 ft) high relative of the yellow-flowered weed familiar to British gardeners, towers like a tree over the otherwise low vegetation. At this altitude, ground temperatures often drop below freezing at night, but as soon as the sun rises, they may leap to 26° C (80° F) in a few minutes. The giant groundsel has evolved its own method of coping with these rapidly changing temperatures. Each plant has a single stem, crowned by a rosette of leathery leaves, and as new leaves are produced, the older ones die back to form a thick jacket around the stem which conserves heat. In addition to this, the living leaves protect the tender buds by folding over them at night, and rainwater that collects in pools at the leaf bases is prevented from freezing and damaging the buds by a natural 'anti-freeze' which the plant produces. On the open moorland these little pools of slimy water are permanent reservoirs in which insects can breed, and these insects provide food for iridescent sunbirds which flit from plant to plant.

The giant groundsel is a source of food and shelter to many other animals that live above the tree-line. As well as harbouring birds and insects, the plants provide cover for the rock hyrax in bad weather. This dumpy animal, which can be found

OPPOSITE *Giant groundsel*

BELOW *Rock hyrax*

LEFT *Forest at 2,450 m (8,000 ft), Mount Kenya*
BELOW *Black rhinoceros*
OPPOSITE ABOVE *Warthogs*
OPPOSITE BELOW *Mud coated buffalo*

in stony places all over Kenya, has thick fur that enables it to live high up on the moors in an environment too cold for most other African mammals.

As the streams that make up the Tana begin their descent of the mountain, they pass through a remarkably varied tangle of vegetation. Below the open moorland are forests of pines, eucalyptus and bamboo, and thousands of feet below them, where the tropical heat is fiercest, the riverbanks are lined with spreading acacia trees and elegant palms. The thick blanket of plants on the mountainside binds the soil together, but where there is a gap in the river, perhaps where a tree has fallen or where wood has been cut for fuel, the rains quickly wash the soil away, staining the river a dirty red. At the base of the mountain the once-clear streams combine to form a sluggish warm river that loops eastwards towards the coast – the Tana.

On its long journey down the mountain, the Tana passes through some of the most fertile land in Kenya. The volcanic soil is rich in nutrient minerals, and in many places the original forest has been cut back by farmers to be replaced by acres of tea and coffee bushes, neatly planted in regimented rows. Around the mountain villages, bananas and maize ripen in the strong sunshine. But where the land has been used carelessly, the rain that creates so much abundance can also destroy it through erosion. Where a patch of ground is left without the protection of plants, a brief downpour will wash away the precious top-soil and as the Tana flows on, it becomes more and more turbid.

Away from the farms and plantations, glades in the forest draw grazing animals more commonly found at lower levels. Most of Kenya's rhinoceros live in the hot and dry lowland savanna, but some make their way up the mountain to graze on the thick grass. Some elephants live permanently in the forest, picking their way through the trees with such care that even a sizeable herd is difficult to hear. A muddy slipway at the riverbank shows where they have come down to drink. The tender grass is also ideal for the warthog, which despite its menacing appearance is strictly vegetarian. It often feeds by kneeling down on its forelegs to crop the grass with its sharp teeth. The slightest disturbance will send it trotting briskly for cover.

The combination of warmth and dampness at the river edge is perfect for insects, and they breed here in huge numbers. For grazing animals like the warthog and buffalo, some insects can become an unpleasant problem. As well as being attacked by bloodsucking flies, they are sought out by insects that lay their eggs on mammals – the larvae develop parasitically by eating the living tissue of their unfortunate hosts. But the river itself is a source of relief. Wherever a bend in the river channel has thrown up a bank of silt, warthogs, buffalo and many other grazing animals gather to wallow. This is not just a way of keeping cool. The mud that the Tana washes down is very adhesive, and when dry forms an almost impenetrable jacket around an animal. Until it flakes off, the thick cake of mud will keep insects in check.

For many smaller animals, the insect life of the river edge is a source of food. The number of insects living near the river at low altitudes is many thousands of times greater than on the high moorland, and a bewildering array of animals feeds on them. In the trees and bushes, horned chameleons search for butterflies and crickets, with the aid of their independently swivelling eyes. Once they have spotted their prey they pursue it with a jerky, almost hesitant gait. The chameleon's perfect camouflage and its stealthy movements allow it to approach close enough to its victim to snatch it up with a rapid flick of its sticky muscular tongue before it is even aware of any danger. The agama lizard also hunts insects in the trees and on the ground, but the male, unlike the chameleon, is highly conspicuous, his skin a brilliant metallic red and blue. His vivid coloration is used in combination with an energetic 'press-up' display to attract a mate. Along the banks of the river, the monitor, Africa's largest lizard, scrapes away at the soft mud in search of insect grubs and worms.

When the annual rains begin, the river floods its banks and there is a burst of insect life in the adjacent swamps. As insects take to the air to find food or a mate, predators flock in to take their share. For the insect-eating birds of the Tana River, the weeks preceding the rainy season are filled with preparation for breeding, so that their young will hatch when the food supply is at its best. Bee eaters and sand martins tunnel into the loose earth of the riverbank to lay their eggs out of reach of other animals. The sand martin's tunnel may be up to 90 cm (3 ft) long, ending in a small chamber lined with feathers in which the young develop. Swifts construct their nests under rock ledges, or even under the spans of bridges. The cup-like nest is made of a mixture of feathers glued together with saliva, which the birds shape and then leave to harden: although it is light, it is also very strong, and can hold the parent bird and the growing brood without collapsing.

Of all African birds, the oxpecker has one of the strangest methods of obtaining food. This colourful relative of the starling can be seen clinging to

RIGHT *Agama lizard*

OPPOSITE ABOVE *Three-horned chameleon*

OPPOSITE BELOW *Nile monitor lizard*

the fur of antelopes, warthogs, and even giraffes, probing with its sharp bill for parasitic ticks. The oxpecker is completely dependent on its hosts for food, and even mates while on an animal's back, leaving only to lay its eggs. It is very persistent and thorough in its feeding, and its sharp claws and beak sometimes seem to irritate its host. During the breeding season it will even pull out hairs from the host's hide to line its nest. However, the host animal does benefit from the bird's grooming, and for the most part its scampering and pecking are ignored.

As the Tana reaches the plains below Mount Kenya, its banks take on a very different character. For much of the year, the surrounding land is parched and brown, and the river is a strip-like oasis, where a tall forest grows like a long gallery. From the air the Tana looks like a green ribbon stretching away from the high ground to the west across open savanna. Here the river becomes a

LEFT *Red-billed oxpecker and buffalo flies on a white rhinoceros*

OPPOSITE ABOVE *Weaver birds nests in acacia trees*

BELOW *Golden palm weaver bird*

OPPOSITE BELOW *Layard's black-headed weaver bird*

focus of activity for wildlife. Sparrow-sized weaver birds, probably the commonest birds in Africa, use the riverbanks as a safe nesting site. Male weavers construct their nests from strips of leaves and riverside grasses which are skilfully woven together to form a spherical or pear-shaped chamber, usually attached to the thorny branches of an acacia tree. The weavers prefer branches that overhang water, because nests situated above the river are less accessible to predators. Some species of weaver go even further and build a long tube on to the entrance of the nest, which dangles freely in the air. This tube, which may be over 30 cm (1 ft) long, prevents tree snakes entering the nest and reaching the vulnerable young inside. Each male bird uses his newly made nest as an advertisement to attract a mate. During the breeding season, the riverbank is alive with the movement of the bright-yellow male weavers, who display to passing females by fluttering their wings energetically while hanging from their nests.

During the dry season, the Tana is slow and

shallow. As waterholes and other rivers dry up, animals make their way to this dependable source of water. The wooded banks can easily conceal a lion or a leopard, so animals like antelopes and baboons approach the water with nervous caution. The river itself can conceal danger. Crocodiles lie almost completely submerged near the banks waiting for animals to stray within their reach. With a sharp snap of its jaws the startled victim is caught, and the crocodile then pulls it below the surfacewhere it drowns.

When the water level is low, clusters of brightly coloured butterflies congregate on the mud, probing it with their thread-like proboscees to sip at the mineral-rich moisture. In the deepest pools, the hippos' daytime languor is interrupted by brief but noisy fights as they jostle for space. When evening falls they leave the river to graze, returning before dawn to the safety of the water. The wood stork hunts for fish in the murky water by using its feet to drive its prey from their hiding places under the bank, spreading its wings perhaps as an aid to balance, or in an effort to shade reflections. At this time of the year, much of the riverbed is exposed, and as the rainless weeks pass by, the animals of the river are concentrated in what water remains.

Then, as the river's flow becomes almost imperceptible, clouds build up over the moorland and mountain peaks as a new rainy season begins. Storms swell the Tana's tributaries and within a few days the lowland river is transformed into a surging tide thick with silt. Trees and branches are swept downstream, Egyptian geese and other waterbirds are thrown about in the flow, and the Tana becomes a barrier that not even elephants can cross. Vultures and bald-headed marabou storks soar over the river in search of carcasses of animals that have been drowned in this annual flood, while all along the Tana's length, herons, storks and egrets are feeding their young. Away from the river, the trees and bushes which have been leafless during the last weeks of the dry season become flushed with green.

As the river nears its delta it widens, and the gallery forest becomes thicker. This lowland forest is the home of two very rare species of monkey, the Tana River colobus, and the mangabey. Unfortunately for these animals, the gallery forest is an important source of timber for the pokomo people who live on the lower reaches of the river. As they cut down trees, the monkeys' habitat shrinks and if this forest eventually disappears, the monkeys will vanish with it.

The Tana completes its journey to the sea through miles of mangrove swamps. The mangrove, a shrubby tree with stilt-like roots, is one of the few plants that can survive at this meeting place of fresh and salt water. Over the centuries, its roots bind the river silt into stable banks rich in organic matter that has been brought down-river. At low tide, countless tiny fiddler crabs emerge from their burrows to sift through the silt for food. They are easy prey for crab plovers and other waders which winter here after breeding as far away as Siberia. As the rain-swollen river pours out to sea, the soil brought down from the distant mountains streams out into the blue waters of the Indian Ocean, and smothers the coral reefs that lie offshore.

Life on the Tana River is dependent on this annual flood, and animals breed and plants flower and set seed at exactly the right time to take advantage of the yearly rains. But up-river, something is happening which might cause a lasting change to this pattern of the seasons. After construction work is finished in 1982, the Tana will be filling one of Africa's largest reservoirs. Over 21 km² (8 square miles) of land will be flooded to supply water for Kenya's growing population. One dam has already been built, and where thorn trees once grew in an arid landscape, rice now flourishes on fertile expanses of irrigated soil. The land is very productive, and birds like the cattle egret, which have taken advantage of man's agriculture to spread around the world, have not been slow to move in here. Yet, for the original inhabitants of the river, the dams may spell disaster. Life in the lower river depends on the annual wave of silt which feeds fish and insects and hence many other animals. Where dams are built, the flow of the river is reduced, and the life-giving silt falls to the bottom of the reservoir — for the animals of the delta country, this means the loss of their most important source of food. For thousands of years the Tana's yearly flood has maintained a richness of plants and animals that is unrivalled in this part of Africa, but with the construction of the new dam, much of the wildlife of this river now faces an uncertain future.

L.G.

ABOVE *Red mangrove tree*

RIGHT *Fiddler crab*

maximum 50 ft.

maximum 100 ft.

maximum 28 ft.

4. OCEANS AND ISLANDS

WHALE OF A TANGLE   *producer Richard Brock*

The big whale known to science as *Megaptera novaeangliae* was given the name 'humpback' by nineteenth-century whalers, for as it dives its back curves upwards out of the sea, forming a dark grey hump above the waves. Although less than half the size of its massive relative, the blue whale, the humpback is still a giant of the ocean, measuring up to 15 m (50 ft) long and weighing about 40 tons when fully grown. Seen underwater, the humpback is quite unmistakable since its flippers are a quarter of the length of its body – far longer than in any other whale. When it swims, they move like a pair of slowly flapping wings. The knobbly edge of the flippers is another feature unique to the humpback; the bumps are formed by bones equivalent to the finger-bones of a human hand, a reminder that the humpback is a mammal like ourselves.

The long flippers act as rudders in swimming, giving the humpback great manoeuvrability, and are sometimes used in feeding as well, scooping prey towards the whale's cavernous mouth. That mouth can take in several tons of water and food at a single gulp, the water being forced out again through the baleen plates whose tangled fibres filter out the food. The tongue then scrapes the food off the baleen so that it can be swallowed, and the humpback is ready for another huge mouthful.

A humpback's stomach can hold 600 kg (1300 lb) of food and it takes a highly productive feeding ground to satisfy its enormous appetite. Subarctic waters provide such a food source, for although extremely cold, they are rich in nutrients and therefore in the shoals of krill (tiny, shrimp-like creatures) and small fish which form the humpback's staple diet. The whales spend the summer

OPPOSITE *Humpback (top), blue whale (centre), pilot whale (bottom) to scale*

RIGHT *Humpback whale*

months feeding in these icy seas, but must return to the warmth of tropical oceans to give birth to their young, a journey of several thousand miles. They migrate southwards (northwards in the southern hemisphere) swimming slowly and generally keeping fairly close to the shore, which may help them to navigate.

During this period the whales do not eat, and they continue without food for the first five months in the tropics, relying instead on the thick layer of blubber accumulated during the summer months. Soon after they arrive, those females who are pregnant give birth to the young ones they have carried for 10 or 11 months. As in other whales, the young

Humpback whale travels north to Canada

humpback emerges tail-first and is pushed to the surface by its mother to take its first breath. A month or so later, the new breeding season begins and the humpbacks embark on their ecstatic courtship displays, leaping almost clear of the water in magnificent backward flips.

This is when they begin to sing the strange haunting songs that have made the humpback one of the best known of all whales. These songs are the longest calls in the animal kingdom – each song lasts between 6 and 30 minutes, the average being about 15-20 minutes. The ocean echoes with the sound, and divers say that they can actually feel the underwater vibrations pounding against their bodies. Deep bass notes can be heard as much as 80 km (50 miles) away, and are detectable with sensitive instruments at a distance of over 1100 km (700 miles)! Humpback songs are not static like those of birds, but develop from year to year, as

these intelligent animals improvise on the original 'tune'. The song that the humpbacks begin to sing when the breeding season commences is identical to the one they sang at the close of the previous season, but as the months pass they will introduce variations on the song, so that by the end of the breeding season it is noticeably different. In the space of a few years these small changes accumulate, and the song pattern takes on a completely new form. Not surprisingly, the song that one population of whales sings is entirely different from that of another group.

There are thought to be about eleven quite separate populations of humpback whales, five in the northern hemisphere and six in the southern hemisphere. Each has its chosen breeding grounds and their fixed migration routes help to keep them apart. The West Atlantic population breeds in the Caribbean and then moves north, in about June, to hunt for food in the waters off New England and Canada, as far north as the Hudson Strait. Some travel even further, to feeding grounds off the coast of Greenland, but it is the humpbacks that travel to Newfoundland that we decided to film, for they are the centre of a controversy between conservationists and fishermen, whose livelihood the humpback whales threaten.

In coming to Newfoundland, the humpbacks are entering what was once enemy territory, for the people of the coastal villages used to kill whales in great numbers. Their main quarry were the diminutive pilot whales, which were driven into small inlets by men in boats, and then slaughtered. Some areas of shallow water off Newfoundland are strewn with the bones of thousands upon thousands of whales that fell victim to the local whalers. Although it was mainly small whales and dolphins that the coastal whalers captured, large whales, including humpbacks, were occasionally caught by this method. And the humpbacks had a far more dangerous enemy in the whaling ships: being slow-moving and in the habit of staying close to the shore, they were prime targets for such ships in the early decades of this century. Within 50 years their numbers had been reduced to 10,000 or less, from an estimated 100,000 before hunting began. When a species is reduced to such a small population as this, it is in severe danger of extinction and in 1966 the International Whaling Convention gave the humpback total protection.

Since that date, Canadians have been prohibited from killing humpbacks, and in 1973 a law forbidding all whaling in Canadian waters, or by Canadian citizens, was passed, giving the Newfoundland humpbacks double protection. In spite of this, between 10 and 20 humpbacks die every year off the coast of Newfoundland through becoming entangled in fishing nets. The damage they cause to the nets runs into millions of dollars, for which local fishermen, dependent on the sea for their livelihood, receive no compensation. Small wonder then that these men now talk of resuming whaling to protect their only source of income. But if this were allowed to happen it would be a tragedy for the humpback, a species that has only just been rescued from the brink of extinction. In an effort to find a solution that benefits both fishermen and whales, the Whale Research Group of Memorial University, Newfoundland, is engaged in an intensive study of the humpbacks and how they feed.

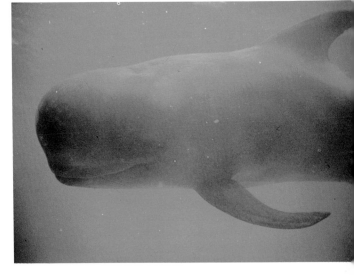

TOP *The long flippers act as rudders in swimming*

ABOVE *Pilot whale*

107

The humpbacks' main quarry in these waters are small fish, scarcely 230 mm (9 in) long, known as capelin. In May and June huge shoals of these fish come close to the shore to spawn in shallow water, and the humpbacks have an ingenious and efficient method of catching them. Diving to a depth of about 15 m (50 ft) below the fish, the humpback spirals upwards beneath them, blowing a continuous stream of bubbles as it goes. The bubbles appear on the surface as a circle about 3 m (10 ft) in diameter, within which the sea seems to be boiling furiously. As the bubbles rise, they carry the small fish with them, concentrating the shoal in a dense layer at the surface of the sea. The humpback then appears in the middle of its circle of bubbles and gulps down great mouthfuls of its tightly packed prey. Sometimes a pair of humpbacks will cooperate in this fishing enterprise, and between them create a much larger bubble raft, up to 30 m (100 ft0 in diameter.

The humpbacks' unique fishing technique is difficult to make out at sea level, but from the Newfoundland cliffs, researchers have a perfect view of the whales in pursuit of their prey. They can build up a very accurate picture of the humpbacks' feeding behaviour, and record the precise activities of each whale. Every humpback has a pattern of black and white on its tail flukes which is as characteristic as a fingerprint, and from this it has been found that the same whales appear near the shore year after year.

The whales reap an easy harvest among the capelin, but unfortunately it is a harvest that man has an interest in too. Along the Newfoundland coast fishing nets are set to trap the capelin as they make their way towards the shore. In the past, the capelin were also fished out in the open sea, by large fishing vessels from other countries as well as by local boats, but the shoals have become so depleted that this is no longer profitable. The drastic decline in numbers is not surprising, for the capelin are caught before they can spawn, when the females are still swollen with roe, so that future generations are destroyed along with the present one. And twice as many fish must be caught as are sold, since the males are not particularly good to eat – these are picked out and dumped as waste, while the females are frozen for export to Japan.

As the declining numbers of capelin have brought the fishing operation inshore, it has, unfortunately, done the same for the whales. With fewer fish out at sea, the humpbacks now pursue

OPPOSITE *Humpback tail flukes*

BELOW *Humpback vents air*

the capelin further and further as they swim towards the beach, following them almost into the breakers – and, all too often, into the fishermen's nets. The humpbacks become caught in the nets, and in their desperate efforts to free themselves, may become even more firmly entangled. Since they are mammals, they must come to the surface regularly to breathe, and if the net prevents them from doing so, they drown. And the owner of the nets, forbidden from killing the whale, must watch helplessly as the struggling animal destroys over 10,000 dollars' worth of gear.

It is here that the Memorial University research teams steps in, for as well as studying whales they run an emergency rescue service for those trapped in nets. A telephone call from the fisherman to the Whale Research Group brings the team to the spot within a few hours, and they then set to work to disentangle the captive humpback. All their skill and experience is required to do this without panicking the animal: if that happens its frantic thrashing can make matters even worse.

In one season the rescue team answers up to 150 calls, and generally they are successful in freeing the whale, but even so there are losses for the fisherman. The catch that was in the net is likely to be lost as the whale is released, and any holes it has made must be repaired. This can take a week, and with the capelin season only two weeks long, the fisherman's annual income is halved as a result.

The rescue service is only a short-term measure and if the researchers are to head off the fishermen's demands for the right to kill whales, they must produce a more effective solution to the problem. One possibility they are exploring is the use of audible warning signals to alert the whales to the presence of the nets. Three types of device that can be attached to the nets are now being tried out – clangers, beepers and pingers. The clanger consists of two metal cylinders on a piece of rope, each cylinder being attached to a buoy. The two buoys are of different size, and as they move up and down with the waves they make the cylinders clang together continuously. The beeper is a more complicated device, and more expensive since it runs on batteries, but it emits a piercing beep which carries well through the water. The third device, the pinger, is also battery-powered and makes a high-pitched sound that the human ear cannot detect, but is well within the range of a whale's ear. Tests on the effectiveness of these warning devices have only recently begun, but it does seem that the whales are learning to associate the sounds they emit with danger and are keeping their distance. Fortunately, too, the capelin are not warned off by the signals and continue to swim into the nets in the same numbers as before.

Another approach to the problem is to try to make the humpbacks of direct benefit to the Newfoundland villages by bringing in whale-watching tourists. Such is the appeal of whales that people flock to see them in Hawaii, one of the humpback's breeding grounds, and along the coast of California, where the grey whale breeds. New-foundland does not have such a welcoming climate as California or Hawaii, but it has already begun to attract hardier tourists who are willing to pay for a chance to watch the humpback at close quarters from a small boat. If more can be encouraged to come, these tourists could represent a much-needed source of income for the fishing communities. It would be a happy change of circumstance if, after so many years of slaughter, man's new-found interest in these magnificent creatures could pay to safeguard their future.

L.G.

SHIPWRECK *producer Michael Salisbury*

Our small diving boat bounced energetically on the choppy surface of Whitesands Bay in south Cornwall. Any warmth the bright May sunshine should have provided was being whisked away by a fresh breeze cutting across the entrance to Plymouth Sound and over the green hump of Rame Head about a mile to the east. As the bitterly cold water crept its way under my wetsuit it was difficult to suppress a momentary regret that we had not set this film, like most other underwater films, in some warm, tropical paradise. Wildlife cameraman Rodger Jackman, who had also just taken the plunge into the icy dark-grey Cornish sea, looked apprehensive enough to be suffering the same doubts. But then we had both agreed earlier in the

year that the main reason for making 'Shipwreck' was to show that underwater life in the shallow coastal seas around Britain could be every bit as abundant, colourful and varied as that of a tropical coral reef. However, instead of filming this on some anonymous rock we thought it would be more appealing to show how the same plants and animals gradually take over a suitable wreck, turning it into an artificial reef, rich in marine life. The problem had been to find a ship that was still whole enough to be recognizable and which would provide a good range of different places for marine life to grow on or shelter in. Underneath us now in 23 m (75 ft) of water lay what we hoped would prove to be the ideal location: the rusting hulk of the *James Egan Layne*, an American cargo ship torpedoed by the Germans towards the end of World War II.

So that Rodger could concentrate on macro-photography and detailed behaviour, I had asked Peter Scoones, with his assistant Gill Lythgoe, to look after the wide-angle, underwater sequences on location. Peter is a diver and marine photographer of much experience, well-known also for his unstoppable energy and enthusiasm – even now he managed at least a fair impression of relish as we finned downwards, following the line of the anchor chain into the comparative gloom and silence of this cold, murky undersea world.

It is nice when the sense of foreboding that comes with not being able to see where you are going turns into the sort of pleasure you might expect if you stumbled upon a magnificent cathedral on a misty day. The *James Egan Layne* loomed into vision like just such a building, only here we were drifting by at roof level: the ship's top deck and rails, decorated by a gently swaying mass of different seaweeds. Seaweeds grow profusely only on these upper parts, a depth of about 6 m (20 ft), because like land-based plants they need sufficient light to photosynthesize their food.

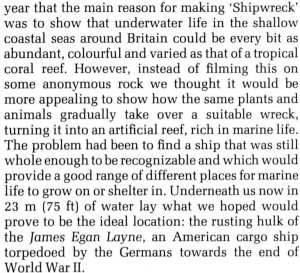

LEFT *The bows of the* James Egan Layne *take shape through the murky sea*

OPPOSITE *Carpets of jewel anemones (Corynactis viridis) adorn parts of the wreck*

It did not take long to realize that this kelp forest could probably provide enough interesting material for a film on its own without venturing elsewhere on the wreck. As we pushed through the tough, leathery stems, fish of the wrasse family, often known as 'wreck fish', flitted around us defending their breeding territories: goldsinny, corkwing and, most colourful of all, a bright-blue male cuckoo wrasse in full breeding colours. An enormous pollack kept station alongside, watchful but seemingly unafraid, while a large shoal of bib turned in a silver flash of unison and disappeared through an opening into the gaping hold below. We stopped to watch a pipe fish nicely camouflaged among the stems. They often swim upright, as this one did, letting the current move them back and forth, so that only the eye, set at the back of a long snout, and the small pulsating dorsal fin show that they are not just floating pieces of waterlogged stick. As we drifted over a barnacle-encrusted winch that stood above the seaweed canopy, a slight movement in the end of a broken pipe caught my eye. A closer look revealed the head of a dumpy little fish called a tompot blenny, a species that have strange, flowery tentacles above their eyes giving them a permanently offended expression. Four more pipes opened within this small area, each one home for a tompot: an encouraging example of the way in which the wreck could provide an alternative to an animal's natural requirements – in this case a crevice in the rock. It was a theme we were anxious to illustrate.

Peter Scoones looked pleased with his last shot, a gliding track over the seaweed to end close to some beautiful specimens whose wide, brown fronds were almost totally covered by a white, furry-looking growth of tiny colonial animals: the moss-like hydroid *Obelia* and the low, spreading sea-mat *Membranipora*. It showed perfectly how larger plants and animals, once established on the wreck, can provide yet more surfaces for smaller ones to colonize. Another example lay close at hand where a large bunch of mussel shells had taken over the remains of an air vent. On the backs of the shells grew a selection of barnacles, anemones, hydroids, sea squirts and tiny tube worms. Intense competition for living space like this would obviously be another important subject for the film.

Anxious to explore further we swam through the bow rails and down the cliff-like sides of the ship. The bottom lay another 18 m (60 ft) or so below, although there was no sign of it in the deep gloom. Because less light falls on these rusting sides, seaweeds give way to an assortment of static, plant-like animals, some with feathery arms like the sea firs, batteries of stinging polyps like the beautiful sea fans or sticky tentacles like the anemones, all designed to catch food from the passing current. The whole area looked more like a carefully planted garden as we drifted downwards past fluffy beds of plumose anemones in pink and cream and by borders of sprouting white soft coral, ghoulishly known as 'dead man's fingers', until at a depth of 23 m (75 ft) we finally touched down on the sea bed.

Stretching away from the *James Egan Layne* for miles in every direction lies this featureless and seemingly barren plain of sand and broken shells. There *are* animals that have developed ways of coping with the shifting particles and we would need to discover and film them, but compared to the flamboyant variety of species on and around the wreck, the sea bed here is a desert.

Like an underwater oasis, the ship towers towards the pale green of the surface. Now that the fish were silhouetted against the light as they swam back and forth through holes in the side, it was possible to appreciate the prolific numbers attracted to the wreck for food and shelter. An enormous shoal of horse mackerel lazily swirled away from us as we headed through one of these gaps into the dark belly of the ship. It was at this spot, where sand spills into the lower hold, that on a later dive I had noticed the grotesque outline of a large angler fish lying motionless on the bottom, camouflaged by blotchy flaps of skin. The angler fish is so named because of the 'rod' on the end of its nose tipped with a fleshy lure. This it waves about to attract other fish near enough to be gulped into the outsize mouth. Rodger spent many hours patiently watching an angler which we had transferred to his large sea-water tank, before he managed to record this bizarre event on film. Another quite common but difficult animal to film was the cuttlefish – particularly as we wanted to show how this big-brained hunter stalks and captures its prey. On the other hand the glinting silver shoals of bib and whiting that swam among the wartime debris strewn over the floors of the holds seemed unconcerned by our presence and would be relatively easy to film.

The underwater cathedral image again came to mind as we finned upwards through the cavernous gloom of the 'nave' towards where the bare ribs of a broken section of deck let in shafts of rippling light. It looked beautiful to the eye, but filming in here would be a real problem. Bring in lights and most of the fish swim off; leave them on the surface and nine out of ten days the water is so murky and natural light levels so low that you would not even get an exposure. The snag is that you can never tell what the underwater conditions will be like from the shore, which as it turned out meant organizing visit after visit to the wreck until by sheer persistence Peter hit upon a couple of days when visibility was up to 8 m (26 ft) or more – enough to take the wide angle shots. When visibility proved to be poor, it was a case of using lights but restricting filming to closer shots of the 'static' plants and animals.

Back on the surface after that first exploratory dive, it was a relief to feel such confidence in the choice of location. In practice the *James Egan Layne* certainly provided a rich variety of subjects to film on a fascinating backdrop. Much of the behaviour and some of the detailed work on smaller species Rodger Jackman filmed in his marine studio in Paignton. This meant building a special outdoor sea water holding tank near the harbour in which to keep the many bits of wreckage and the animals collected during a number of dives. If as the result of all this effort the finished film did not give a good idea of the staggering complexity, colour and beauty of marine life on a British reef, then the failure is ours because if you can cope with the cold and the often poor visibility, it is all there to see not far from any holiday beach.

Furry-looking growths of tiny colonial animals often cover the kelp fronds on the upper deck. In the foreground the moss-like hydroid Obelia *and behind the mosaic structure of the sea-mat* Membranipora. *These 'static' animals catch food from the passing current*

ABOVE Peter Scoones films the densely packed animal life that encrusts the steep sides of the James Egan Layne. A sea-fan Gorgonacea stands out in the centre of the lit area

RIGHT Fierce competition for living space beautifully illustrated by this crowded selection of animals. Pink plumrose anemones grow on a bed of mussel shells, themselves surrounded by red sea-squits and 'dead man's fingers'

OPPOSITE ABOVE Corkwing wrasse (Crenilabus melops) like this one are among several species of colourful wrasse common on the ship. They are sometimes known as 'wreck fish'

OPPOSITE BELOW A female cuckoo wrasse (Labrus mixtus) swims among some rusting debris. All cuckoo wrasse start life as bright orange females, some individuals turning into blue-striped males after a few years

BELOW *Underwater life in the shallow seas around Britain can be as colourful and varied as that of a tropical coral reef*

ABOVE Tubularia indivisa – an attractive hydroid that grows profusely on the wreck. Similar marine animals look like plants, with branching fans or tentacles to trap their food

LEFT Alcyonium digitatum, a soft coral that comes in orange, white, pink and yellow and is ghoulishly known as 'dead man's fingers'. Here the stinging polyps are extended for feeding giving the 'fingers' a furry outline. Under the central, orange dead man's finger, an adult male lump sucker or sea-hen (Cyclopterus lumpus)

ABOVE *A father lasher or short-spined sea-scorpion (Myoxocephalus scorpius) illustrates near perfection camouflage. These fish are so difficult to spot that it is quite possible unknowingly to put your hand on one. A sudden movement and a sharp dorsal spine through your diving glove will be the reward!*

LEFT *A colony of light-bulb tunicates (Clavelina lepadiformis). These animals do not move about but filter their food from the sea water sucked in through the large opening on top and blown out of the small one at the side*

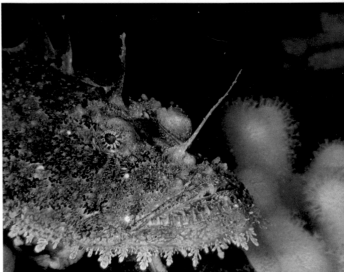

ABOVE *The cuttlefish (Sepia officinalis) is one of the James Egan Layne's most sophisticated hunters – with a large brain and very sharp eyes similar in design to our own. The cuttle shell, which often ends up in budgerigar cages, stiffens the sac-like body from the inside and the swimming frill, combined with a water jet, helps the cuttlefish manoeuvre with great precision backwards or forwards. They also possess a mechanism for changing the patterns and colour of their skins at a speed unequalled in the animal world*

OPPOSITE *Slower-moving fish like these bib (Trisopterus luscus) find shelter and food in the cargo holds*

ABOVE RIGHT *A cuttlefish, now in its camouflaged colouring, pounces on a shrimp. The two middle tentacles shoot out beyond the others to grab the prey.*

ABOVE *An angler fish (Lophias piscatorius) twitches its 'rod' (a modified dorsal ray attached to the top lip) to attract small fish. If one comes near enough, the huge mouth will spring open sucking the unfortunate prey inside*

OPPOSITE ABOVE *A bed of brittle stars (Ophiothrix fragilis) hold their arms into the current awaiting the arrival of suitable food*

FAR LEFT *A typical place to find a lobster (Homarus gammarus), in this case being shared by a group of brittle stars. The James Egan Layne provides many such spots, under broken pieces of wreckage, where lobsters like to shelter*

LEFT *Many different types of sea-slugs graze upon the firmly attached animals that encrust the wreck – this one is called Coryphella. The white tips of each appendage along the slug's back contain stinging cells incorporated from other animals that it has eaten – a second-hand defence system*

ABOVE *Starfish and sea urchins are common predators and grazers in British waters. Like the sunstar (Solaster papposus)*

RIGHT *Underwater plants do not escape from being eaten. Here a blue-rayed limpet (Patina pellucida) munches away at a kelp frond*

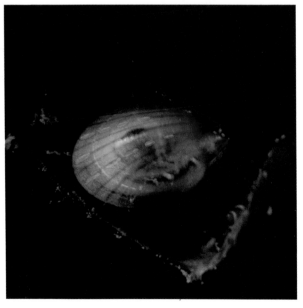

An Interview with Maurice Tibbles, the cameraman who filmed

THE BIRD THAT BEAT THE US NAVY *producer Maurice Tibbles*

Maurice Tibbles

Maurice Tibbles is a naturalist and himself belongs to a very rare species. He is a wildlife cameraman, one of an internationally known team who worked on 'Life on Earth', 'The Voyage of Charles Darwin' and 'Wildlife on One'. His is a small and exclusive profession, highly skilled and, at times, rather dangerous. But Tibbles lets his reputation and craft speak for him. Quiet, unassuming and gentle, he is the opposite of what one would expect from an ex-Fleet Street photographer who spent 14 years on the *Daily Mirror,* photographed all the Royal Family, flew with the RAF's Red Arrows display team, went on tour with the Beatles and to war in Indonesia and Borneo. Asked for his favourite cutting from his press book, he typically shows most pride not in any of the glamorous Street-of-Shame assignments, but in a centre spread of waders in flight.

He was born in 1935, his father a gardener, his mother a cook in a large house, and his childhood habitat a little Gloucestershire village called Coates not far from where Prince Charles now has his house. These rural surroundings in the West of England stamped themselves on Tibbles's life very early on. Television was in its infancy; children made their own pleasures. Most boys played cricket, and most boys collected birds' eggs. Tibbles was good at cricket – captain of his school – but unlike the other boys his interest in birds did not disappear once he had collected a boxful of coloured eggs. Natural history became, as he puts it, an 'obsession'.

There is always one boy in every country school with a frog in his pocket. In his school Tibbles was that boy. He spent days watching moorhens and building dens to spy on owls in barns. His friends thought he was odd when he did not want to come and kick a tin can around under a streetlight. But at the age of nine, he showed them up: he wrote a book on the natural history of the village pond. This was later published as a school book, with his own illustrations.

But for all his interest, there was no way at the time for him to earn a living from natural history. So, when Tibbles left Cirencester Deer Park School at the age of 15, he chose the next best thing and became a trainee gamekeeper. This lasted eighteen months. He loved breeding pheasants as it gave him a lot of time to be out in the wild, but with his village upbringing he was too good when it came to shooting them. So he left and, at the age of 17, joined the Royal Air Force.

On leaving the RAF and returning to England, he became a photographer with the Peterborough *Evening Telegraph.* By then he had already owned a Kodak Retina camera, as well as a Rolleiflex he had bought at a duty-free port. But now, interested in news, Tibbles decided that if he was going to be a photographer, he was also going to look like one, and bought himself a big pressman's camera.

Tibbles must have had more than the right-looking camera, for it was only a couple of years later that he was taken on by the London *Daily Mirror,* at the incredibly young age of 23. Most of

Cuckoo laying its egg in a reed warbler's nest

Award-winning Daily Mirror *photograph*

those around him had worked their way to Fleet Street, with a decade or so on local papers, and Tibbles felt out of his depth. When he went to a London photocall all the cream of the Fleet Street photographers were there and Tibbles, with a very different background, was terrified. But he noticed one thing. For all their experience, the other photographers tended to take the same picture. He was so naïve, he would often hang around after the other others had left and find that he could then take the most interesting photo. The difference, he explains, is that he was a photographer who worked on newspapers – and most of the others were newspapermen who took pictures.

So it is that he is proudest of his newspaper spread showing nature's equivalent of the rush hour, and of a photo – two years in preparation and eleven seconds in the taking – of a cuckoo laying its egg in a reed warbler's nest – a photograph

taken when filming the BBC film 'The Private Life of the Cuckoo'. Years of planning and reading were needed to capture that unique moment – after all, the cuckoo could have laid its egg anywhere. Tibbles had learned enough about the bird to predict how it would behave. And appropriately enough, when Tibbles became *Mirror* Photographer of the Year, it was for a photograph of a black hand and a white hand holding a dove that had hatched both a black and a white chick on Christmas day. At first, no-one believed the story and could only be convinced of the truth on hearing it from the head of the zoo.

While reporting in the Far East, Tibbles contracted scrub typhus, and became so ill that the RAF had to fly him back to England. It was at that point that he realized he should choose to do what he *wanted* to do in life. During the time still photography was his job, cine photography had

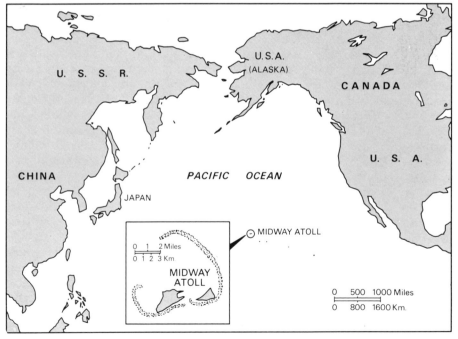

Midway Atoll

become his hobby; Tibbles shot his first piece of wildlife film in 1961. It was in black-and-white and was bought by the Survival Unit for one of its first programmes. Tibbles also made a film on the lapwing, working at weekends and on free days. In 1972 he was offered a job as director-cameraman with the BBC's Natural History Unit in Bristol.

Wildlife filming is unlike any other kind. One example of this is the size of the team; the larger the film crew, the less natural behaviour is going to be caught on the film, so Tibbles mainly works by himself or with a single assistant. He uses similar cameras to other documentary cameramen, but with a wider range of lenses. Tibbles remembers making a film in Africa that needed footage of termites the size of a pinhead. That sort of filming is normally done under a microscope in controlled conditions, but Tibbles had to make do and adopt equipment that had served a week before for filming elephants! Special briefs are the norm. For a film on rabbits, Tibbles built a burrow that would allow him to film a tracking shot underground behind a rabbit, unfilmable in any normal way. Very often he will run four or five different projects alongside one another, waiting for the right conditions to film any one of them.

Tibbles accepts that there are advantages in

growing up a country boy and then learning the tricks of the pressman's trade. Of course, before making a nature film, Tibbles reads extensively and becomes a mini-expert on the particular subject. However, he is no trained academic. What makes up for this – and indeed supplies his special magic – is his background. Half the battle in wildlife filming is anticipation. So that the camera is running in time, it is vital to know the signals that lead up to behaviour such as mating.

When something happens, the picture has to be grabbed then and there, and no retakes. He recalls that one of his first press pictures was of Winston Churchill feeding deer in Richmond Park. Being young and green, Tibbles asked him to pose. Churchill carried on: 'You must learn to take your pictures on the wing, my boy.'

A most successful example of this was to be Tibbles's film on the gooney bird 'The Bird that Beat the US Navy'. This was made on Midway Island, a tiny coral atoll in the middle of the Pacific Ocean. Normally peaceful, Midway had in 1942 been the scene of one of the fiercest battles with the Japanese navy. The battle of Midway changed the whole course of the war in the Pacific, and the island is still a vital US base.

When Tibbles saw the place, his first impression

Courtship dance

LEFT *The island mascot*

OPPOSITE ABOVE *Laysan albatross with its 1.8 m (6 ft) wingspan*

RIGHT *With young*

was one of a gigantic bird cage, full of Disney birds. The entire surface of the island was covered in thousands upon thousands of big, comic birds. And when he saw that the same bird had been adopted as the island mascot by the officers on Midway, with badges, T-shirts and soft toys in its image, Tibbles recognized 'a darn good story, with all the elements of humour and action'. And so it was to turn out. Shot over a period of five weeks, the film became the first 'Wildlife on One' – and the first of many programmes aimed at a popular audience and set to achieve high audience ratings.

In the air – and officially – the birds are known as the Laysan albatross, elegant beasts of the sky with 1.8 m (6 ft) wingspans. But once they have landed, often clumsily, too quickly or on their noses, they are known as gooneys, plump clowns standing some 90 cm (3 ft) high. The gooney is a a rare bird, nesting only along this northwestern chain of Hawaiian islands, where food sources are right for the young. Midway is now shared between tens of thousands of gooneys, and thousands of US naval personnel. And thereby hangs the tale. It is not just

an important island historically; situated almost exactly between Vladivostok on the east coast of Asia and Los Angeles on the west coast of America, it is a satellite airfield that could still be strategically decisive. But the Russian ships monitoring the activities on Midway proved a far smaller hazard for the Americans than 4 kg (9 lb) gooneys being sucked into the engine intakes of the big jets that use the island. At one time, there were nearly 1000 aircraft 'bird-strikes' in two years on Midway, more than anywhere else in the world. To move the birds off what was now the navy's island, the Americans burned flares, fired bazookas, filled dustbins full of nails to blow them up and had the Marines in to mine them. Some gooneys died, a few moved off . . . and more came back to breed. It was still the 'gooney's' island.

Now the gooneys live happily with the navy, because the navy has given up! Every person about to be stationed on Midway is equipped with a gooney pre-pack of photos of the birds and the island, so that the sailors know the kind of thing to expect from the persistent gooneys. Once, when a chapel was built on a particular gooney's nest patch, the bird rebuilt its nest 'in' the church, on the same spot as before. Midway has the only game warden on the payroll of the US Navy, just to look

after the gooneys! The airfield is the only area where the gooneys and man are still in conflict. Before take-off, the ground crew go around picking up the gooneys sitting on the runway and drive them away in the back of a truck. But on the roads, special slow speed signs are enforced, as the gooneys appear indifferent to their own safety and will not budge.

All through the night, the whole island is filled with the gooney call: a clapping of the bills, followed by a howl. The Hitchcock-like density of the creatures takes its psychological toll. Tibbles tells of an officer who went mad and was arrested after having stabbed and killed about 30 gooneys with a knife.

Most people, however, love the gooneys. Indeed, the impression the island gives is that the gooney is 'all' that is cared about! When the first gooney returns to find a mate – sometimes from as far away as 6400 km (4000 miles) – the owners of a nearby house give a barbecue to celebrate the return of 'their' gooney. And as the birds can live up to 50 years, the loyalty to 'your' gooney is strong. The children do not have teddy bears; they have cuddly gooney toys. Midway must surely have the only golf course in the world which officially recognizes an albatross as a golfing hazard; the golf balls

OPPOSITIVE ABOVE *James M. Bradley, Game Warden*

OPPOSITE BELOW *Fairy tern chicks*

RIGHT *The airfield where the gooneys and man are still in conflict*

ABOVE *Gooney courtship display*

LEFT *Mobbed by sooty terns*

OPPOSITE *An adolescent gooney*

history work.' Tibbles has contributed to several films made in towns, including one called 'The Queen's Garden' about natural history in the grounds of Buckingham Palace, but claims even the thought of driving into London will produce a cold sweat! He is afraid of the world turning into a concrete jungle, where people will satisfy the call of the wild by watching old nature films on television. He is amused that there is argument over whether to put a road through farmland or through Dartmoor, when farmland – which is man-made anyway – may be unnecessary in a few hundred years' time. 'Once you've chomped through Dartmoor, you can never put it back.' He is proud of his contribution to a programme about Kenyan elephant poaching which contributed to a ban on sales of ivory. In 1974, he filmed a bird called the Oo-Oo Ah-Ah, of which there were then only two left. It is now extinct.

His main purpose, of course, is to entertain. He recognizes the need to grab the audience. It is this showmanship that has led wildlife cameramen like Tibbles to search for ways to capture some piece of behaviour for the first time, or to get even more dramatic footage of it. This explains the need for that suspiciously regarded technique of the nature film, the artificial or controlled set-up. There is no way of filming, say, the birth of a wild rabbit, so a set is built and the same piece of behaviour is filmed under controlled conditons. In this way cameramen can get quality pictures without disturbing a wild situation. Tibbles is aware of the dangers of nature films, in that they may sometimes convey an idealized image of wildlife, leading tourists in Tanzania to complain that not enough lions are jumping on zebras.

In 1979, the British Academy of Film and Television Arts awarded Tibbles the title of Film Cameraman of the Year, its equivalent of an American Oscar. When asked for the greatest moments of his life, this was his reply: 'To sit in a hide and watch a barn owl come in, to be that close and know it's a privilege. When I filmed a peregrine falcon for the BBC and she brought a grouse in for the first time to eat, I almost didn't start the camera, the adrenolin was flowing so fast.' His gift is to communicate this love and understanding of nature to millions of people all over the world.

supplied by that club have the same bird stamped on them: the emblem of Midway, the gooney bird, the bird that beat the US Navy.

Not all Tibbles's films come from exotic locations like Midway. When asked to make another 'Wildlife on One', Tibbles jokingly said it would have to be on a subject he could fit in at home between trips, like goldfish. The BBC took him seriously, and another fascinating film resulted. He speaks enthusiastically of casing a microphone in rubber, dropping it into the bowl and annoying the goldfish. 'It screamed', says Tibbles, 'like a cat.' Then he does an impression of the howl of an upset goldfish.

He lives on 7.5 ha (18 acres) in Devon, in what he calls his wildlife studio. Set in *Tarka the Otter* country, his land has foxes, barn owls, a lake and all manner of wild animals. As he says, 'You have to live close to nature if you're going to do natural

Sebastian Cody

GANNETS GALORE *producer Lancelot Tickell*

The Atlantic gannet *Sula bassana* is the largest of our British seabirds and the most spectacular. Its recent history is in refreshing contrast to the many sad tales of near extinction that we hear so often these days. From the turn of this century gannet numbers have been increasing over our seas and spreading to new breeding locations around our islands; this trend continues at a rate of three per cent per year. We have the largest share of the gannet population and the largest single gannetry

(over 50,000 nests) in the world. All around the coastline of the British Isles there are, indeed, Gannets Galore.

Gannets are cliff nesters and nowhere are they seen in more dramatic profusion than on those great spikes of rock, Stac Lee and Stac an Armin, that pierce the ocean along the western cliffs of Boreray at St Kilda where I had always imagined opening the gannet story; it is a place of pilgrimage for all who are fascinated by seabirds and have a feeling for islands. I have an abiding memory of cameraman Huge Miles filming in the distance along the Boreray cliffs on top of a buttress hundreds of feet above the sea; he is silhouetted against the great white stac oblivious of the abyss below. The immense scale of such scenes defies transformation to the small screen and the editor's waste bag claimed much of the wide-angle atmosphere of the place.

At places like St Kilda you have to be a climber to get among gannets, but there are other islands where the birds have occupied all the available ledges and taken over the slopes and flatter areas above. This is where both ornithologist and cameraman can best get an intimate glimpse of gannet family life. The Bass Rock in the Firth of Forth is a place of Scottish history, an ancient hermitage, fortress, prison, and now a lighthouse. For hundreds of years the pageant of man's faith, hope and cruelty have been played out here against a backdrop of gannets, dazzling white on their guano-plastered rocks. An ancient description reads: 'There is nothing in this rock which is not full of admiration and wonder; therein also is a great store of Solan goose . . .'

Solan goose (or variants of that spelling) is an ancient name for the gannet still widely used around the north of Scotland. 'Goose' of course is inappropriate because it isn't one, but solan for me is evocative of the northern isles where my Shetland friends always use it.

The Bass gannetry meets you long before your boat nears the great monolith; a pungent smell drifts downwind which is accompanied by a great

LEFT *'The nest site is the focus of the gannet's existence'*

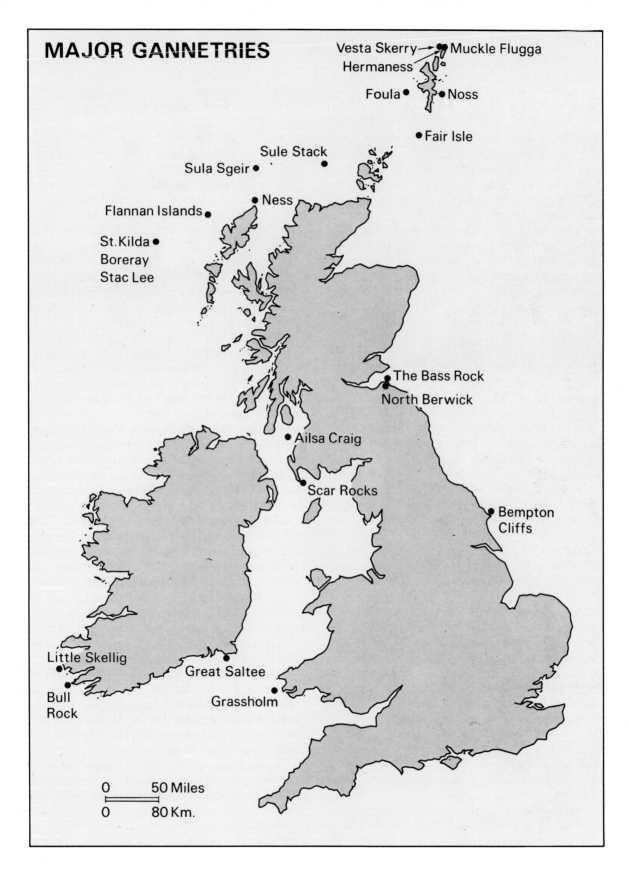

MAJOR GANNETRIES

Vesta Skerry → ● Muckle Flugga
Hermaness

Foula ● ● Noss

● Fair Isle

Sule Stack
Sula Sgeir ● ●

● Ness
Flannan Islands ●

St.Kilda ●
Boreray
Stac Lee

● The Bass Rock
● North Berwick

● Ailsa Craig

● Scar Rocks

● Bempton
Cliffs

Little Skellig
●
● Great Saltee
Bull
Rock ●
● Grassholm

0   50 Miles
0   80 Km.

cacophony raised by thousands of strident voices, lifting wave-like as excitement waxes and wanes. The rock is a powerful attraction to its gannets and although more or less empty during the winter, many are not far away in the North Sea and return promptly with the first sign of spring. Soon the gannetry is crowded with owner-occupiers confidently flying in and vociferously announcing their presence at well-remembered sites, which most will have occupied for many years. Any incomer who does not respond appropriately to that conspicuous bowing signal is in for a nasty reception; confident owners do not hold back from stabbing vigorously. The surprised response to such a welcome should normally be to topple backwards off an exposed ledge and fall away in a ruffled glide, but in the crowded galleries easy escape is often not possible; the attacker persists and may be joined by neighbours from adjacent nests. Devastating and frequently bloody combat is characteristic of gannets and unusual among birds, but it does not make it easy for a female to get to her mate. An elaborate mutual greeting ritual is needed with the female taking many blows on her turned-away head before a pair can share a nest in peace. Once there though, females are as vigorous in defence as males.

A year before we started filming I flew over the Welsh island of Grassholm taking vertical photographs of the gannetry from which to count nests.

You can do that at Grassholm because the cliffs are low and most of the birds nest on slopes visible from above. What strikes you, looking through an open door at 275 m (900 ft), is the regularity of nest spacing, row upon row all the same distance apart. On the ground it is easy to tell that the separation is the minimum possible, just the extent of two outstretched beaks.

Forty-five years ago the late Sir Julian Huxley was so impressed by the spectacle of Grassholm that he produced and narrated Sir Alexander Korda's cinema film *The Private Life of the Gannets*. It won a Hollywood Oscar for the best short film of 1937. At that time the Grassholm gannetry had been growing steadily for over 30 years and numbered about 6000 to 7000 nests. However impressive it was then, it is more so now with over 20,000. The spot where Huxley's camera was placed has long since been taken over and is now deep in the gannetry.

The shoals of herring, mackerel, sprats and similar fish that occur in the Atlantic and North Sea are abundant sources of food for any seabird that can get at them; but they swim deep enough to be out of reach of most surface feeders. Gannets have evolved all the structures and behaviour to exploit this niche and we see it all in their accomplished plunge-diving. A swiftly passing gannet can glimpse a fish below and instantly go for it before continuing on its way; but the performance

OPPOSITE LEFT *The Bass Rock gannetry*

OPPOSITE RIGHT *'Biting the nape of the neck is an important kind of sexual stimulation'*

RIGHT *Grassholm gannetry*

BELOW LEFT *'The gannets wings are almost 2 m (7 ft) across'*

BELOW RIGHT *'Birds on their are draped and garlanded about the cliffs'*

is best seen when companies of gannets are diving into a shoal. Urgently back-flapping to adjust aim and sometimes calling to warn others, they suddenly plummet towards the surface. In the instant before hitting the water the wings are streamed, arrow-like, and the bird slides out of sight.

Birds spend most of their lives working against the force of gravity but in diving, gannets use it to accelerate themselves and provide maximum penetration of the water at a dive velocity of up to 100 km/h (60 mph). The impact with water could be damaging and impede penetration to depth, so the birds are perfectly streamlined and cushioned by air sacs to minimize shock and ensure smooth entry. Underwater they swim with half-closed wings and we can still only imagine what it is like down there among the confused shoals of fish.

I had wanted to a shoot a diving sequence from above and below water. Speculation beforehand had mainly been concerned with how dangerous it might be for the diver/cameraman – visions of all those steely beaks hurtling towards camera! We need not have worried: when Peter Scoones got into the sea off Ailsa Craig, the gannets which had been taking our mackerel with enthusiasm up to then, gave up diving altogether. Obviously, when you think of it, this makes sense: a bird that can see a herring clearly is not going to miss a wet-suited diver and it is certainly not going to get into the water alongside such an unknown fish – not for all the bait in the Clyde!

Gannets were formerly an important source of

meat and eggs wherever nesting grounds were accessible and in the Hebrides that tradition has persisted. It was only in 1930 that the St Kildans were obliged to give up a way of life dependent upon fulmars, puffins and gannets. It is perhaps due to decline in gannet eating that we owe the present burgeoning population. Today there is only one place in Britain where gannets are killed for food and that is on the remote rock Sula Sgeir, way out in the Atlantic northwest of Cape Wrath. For hundreds of years men from Ness on the isle of Lewis have made an annual expedition to Sula Sgeir to collect young fat gannets – gugas they call them. Today the gannet is a protected species throughout Britain and guga meat may not now be an essential part of the Lewismen's diet. But the gannet population of Sula Sgeir is healthy and the Nature Conservancy Council gives dispensation to the people of Ness to continue an ancient rite.

Shetland was only recently colonized by gannets but this northernmost part of the British Isles is also the home of the great skua, or bonxie as the Shetlanders call it. Bonxies are quite capable of fending for themselves but are not averse to turning pirate. Just out from the cliffs of Herma Ness and Vesta Skerry they patrol the sea approaches for incoming gannets bringing fish to their insatiable young. Once engaged, the pursuit is a display of aerial virtuosity and by no means one-sided in its outcome. Now and again a heavily gorged gannet will succumb to the harassment and throw up a fish or two, but more often than not it either has none or can hold onto a light load until the bonxie tires and breaks off the attack. Filming all this on Bobby Tulloch's boat was another of our lucky breaks; sun and calm sea are not typical of Shetland. For Hugh it was hand-held camera all day, action at extreme distance and always coming from the least expected quarter.

Everyone who goes to the Bass Rock knows Fred Marr of North Berwick. Three generations of the Marr family have fished lobsters in the Firth of Forth and acted as boatmen. Throughout the summer, boatloads of holidaymakers circle the rock and listen to Fred's well-rounded commentary on its seabirds before landing and stumbling over the tussocks to the edge of the 'gannet city'; it is real life in a big way, and one sight at least that is better than television.

All this has to be after chicks have hatched, one per nest and in powder-puff down. Later they fledge into black-speckled juvenile plumage, ugly ducklings who will take five years to attain the immaculate white of their parents. The first flight form the gannetry is a testing time; for some just a matter of working up enough determination to jump. Others are not so lucky, they run the gauntlet

of stabbing beaks before they can even get to a jumping-off place, and by that time it may seem a positive relief just to go. A young bird may fly a mile or so but it is too heavy and will eventually splash down and not get airborne again until it has worked off some reserve fat. A little time to get accustomed to the sea, but soon enough it must take to the wing again, learn to fish and begin its way south to warmer seas off North Africa.

For me the most memorable spectacle of the Bass gannetry is the massed flight: mostly of young gannets and those that have not yet secured nest sites. They congregate on the fringes of the nesting area – the club – and from time to time all take flight together. The sky is full of birds wheeling over the sea and contouring the cliffs. Others circle on the standing waves created by the rock, or break off to hammer some shoal hidden below the water, the shower of twisting birds and the bomb-spouts of their plunges flashing in the sunlight.

RETURN OF THE SEA EAGLE *producer Richard Brock*

Two hundred years ago, the distinctive silhouette of the white-tailed sea eagle was a familiar sight over the coasts of Ireland and Scotland. With its wingspan of over 2 m (7 ft) – longer than the golden eagle – it was Britain's largest bird of prey. But wherever this magnificent super-predator has bred, from Iceland as far east as Japan, and from Arctic tundra tó the shores of the Mediterranean, it has been constantly persecuted by farmers and fishermen who have always seen it as a competitor for food. In Britain the onslaught against the sea eagle was so effective that by 1916 only one bird remained, leading a lonely existence on the coast of the Isle of Skye. It looked then as if this imposing species was destined to disappear from Britain for ever, but fortunately for the sea eagle, attitudes to birds of prey have changed and they are no longer seen simply as killers that should be eradicated. With the help of conservationists in Norway and Germany, where the eagle still breeds, the white-tailed sea eagle may once again hunt over Britain's northern shores.

The story of the sea eagle's return begins in Norway, which has the largest population of these birds in the world. Along the deeply indented Norwegian coastline five to six hundred pairs of eagles – about three-quarters of the total number in Europe – scour the fjords and mountainsides for food. This is one of the few places where the adult birds, with their handsome brown plumage and white tails, can still be seen in some numbers. But here, as in Britain, the eagle has had to battle for survival against man. In winter hungry eagles used to be seen as a danger to sheep (although there is little evidence that they actually do kill farm animals), and even children were thought to be at risk from their attacks.

Norwegian farmers devised a unique method to catch these massive birds in the days before shotguns made the task easy. A low stone hut was built on open ground, and bait, perhaps the carcass of a lamb or hare, was put out nearby. Under the cover of darkness a man would enter the hut, often in freezing conditions, to await daylight. As the day progressed the bait was gradually pulled close to an opening in the side of the hut, and the eagle's natural wariness would be overcome by its hunger and the prospect of what looked like an easy meal. Once enticed close enough to the hut, the eagle's leg would be seized by its hidden adversary and the bird killed. Old photographs of trappers with the catches show that this method of hunting was very effective, and every year there were fewer and fewer eagles to be lured to their deaths.

As long ago as the sixteenth century there was a bounty paid for killing eagles in Norway, and their black claws were in great demand as a remedy for jaundice. The persecution of these birds increased in the nineteenth century, and in one 23-year period, 61,000 golden and sea eagles were shot. The sea eagle vanished entirely from the southern part of Norway and was severely depleted in the north. But today its reputation as a sheep killer has almost been forgotten, and although some of the older fishermen still destroy their nests every year,

LEFT *Sea eagle in flight* OPPOSITE *'Constantly persecuted by farmers and fishermen'*

most people accept the sea eagle as a harmless neighbour.

In its stronghold in northern Norway, the sea eagle braves the severe winter, staying put at a time when many other birds fly south to warmer climates. Although it lies north of the Arctic Circle, this part of Norway is bathed by the comparatively warm waters of the Gulf Stream, keeping the sea largely free of ice. This is crucial for the eagles because it means that fish can still be caught even in the worst winter months. Other fishing animals, especially the otter, provide scraps that the eagles can feed on during the most difficult time of year. In midwinter the temperature may drop below −20°C (−4°F), the sun never appears and there are only a few hours of twilight during which the eagle can hunt, or look for remains washed up on the shoreline. When food is particularly scarce, the eagles spend hours perched motionless in low bushes to conserve energy, rather than continuing a possibly fruitless search for something to eat.

Although the sea eagle is generally thought of as a fishing bird its diet shows great variety. In Norway, examination of their pellets (indigestible material like fur and feathers that is regurgitated) shows that their main prey is other birds. Eider ducks, puffins, guillemots and gulls are regularly eaten, and sometimes birds as large as swans may be tackled. The eagle's huge wingspan makes it too ungainly to catch birds in flight, so it attacks them on the water before they have had an opportunity to take off. Sometimes pairs of eagles will hunt ducks, picking a single bird out of a flock and harrying it, so that it dives repeatedly. Eventually it becomes exhausted, and then it is easily snatched up from the water by one of the eagles. Sea eagles also feed on rabbits, hares, eggs, fish and carrion of all kinds, including dead sheep and seals, and those living in southern Europe have even been seen eating snakes and tortoises.

The eagles are not only accomplished hunters, but also skilled scavengers and pirates. They are quickly attracted to carcasses, and waste from fishing stations is a frequent source of winter food. Any bird that crosses an eagle's tracks carrying food is likely to be robbed of its meal. Often the eagle's mere presence is enough to make smaller birds like gulls, ravens, and even buzzards drop their food and flee.

It is in fishing, however, that the eagle shows off its real talents. It operates by surprise, snatching fish from the surface with an almost leisurely air. The eagle glides low over the water with its flight feathers spread wide to act as brakes, and the out-stretched talons pointing forwards. As it passes over its victim, the powerful legs swing backwards and the fish is snatched into the air, the eagle's claws scarcely rippling the surface. Its whole body flexes as the fish is dragged from the water, and the weight of the struggling prey hardly disturbs the smooth arc of the eagle's flight. Small fish are eaten on the wing, whereas larger ones are taken to land to be picked apart. Eagles usually fish alone, although a fishing boat or large shoal may attract up to 30 birds wheeling and swooping over the water.

Although Norway's white-tailed sea eagles are enterprising feeders, by contrast with many other birds of prey, such as the osprey, they are un-adventurous travellers. Every year European ospreys travel thousands of miles to winter as far away as tropical Africa, but the adult sea eagles rarely stray far from their territory of about 130 km² (50 square miles). Because sea eagles are so attached to their ancestral homes, there is little likelihood that a pair would cross the North Sea to nest as far away as Britain. While the osprey has re-established itself as a breeding bird in Britain after a long absence, sea eagles have only made

OPPOSITE ABOVE *Chicks eat up to 340 g (12 oz) a day*

OPPOSITE BELOW *and* BELOW *A sea eagle fishing*

OPPOSITE *Some inland
nests are huge*

RIGHT *A high cliff
ledge nest*

brief appearances, to leave again without nesting. Now, however, there is a move to reintroduce the bird to Scotland. Every summer since 1975, sea eagle chicks have made the journey from Norway to Scotland, not by their own efforts, but through the work of The Nature Conservancy Council together with the World Wildlife Fund and the Royal Air Force.

In early spring, Harald Misund, one of a team of Norwegian ornithologists, begins his part in the sea eagle airlift by checking nesting sites for signs of breeding. It is difficult work. Years of persecution have discouraged the eagles from nesting in accessible areas, and many of the nests have to be reached by cross-country skiing, by mountaineering, or by boat. Each pair of eagles has a number of nesting sites in its territory, perhaps as many as ten, but only one is used each year. They may use the same nest for a number of seasons, or they may alternate between favourite sites. Every time they

breed in a nest, they add to it, building a structure of branches, twigs, moss and seaweed that may weigh as much as 225 kg (500 lb). Some inland nests are huge: in one case, over 3000 sticks more than 1.8 m (6 ft) long were found in a nest that was only four years old. But, despite their size, these giant nests, concealed in forests or on high cliff ledges, are fairly inconspicuous. The nests vary depending on their location: on the small islands that lie along the coast of northern Norway, wood is scarce, and here the eagles' nests are just depressions in the ground filled with twigs and moss.

By the time Harald Misund has checked his quota of about 35 pairs, breeding has begun. The first eggs are laid in mid-April; the coastal birds are the earliest to lay. Just over a month later hatching begins, and in the lengthening daylight the male hunts constantly to feed the mother and her brood, usually consisting of two chicks. As the chicks grow, their food requirements – up to 340 g (12 oz) of meat a day – are so great that the female bird must leave the nest to join her mate in hunting. This occurs about two weeks after hatching. Initially the mother carefully feeds her chicks, but after a while food is simply dumped at the nest for them to deal with as best they can.

As soon as the female birds leave their nest, when the chicks are 6-8 weeks old, Harald Misund goes into action, collecting chicks to be flown to Scotland under a special government licence. Despite the size of the black-and-brown fledglings, they are surprisingly unaggressive, and as long as the parents stay away from the nest, collecting them is relatively easy. The young birds ruffle their wings and open and close their beaks at the intruder, but they allow themselves to be picked up without struggling. Only one of each pair of young is taken; under certain conditions, only one bird may survive in each nest, so those removed can represent a natural surplus, and the recruitment of adult eagles in Norway is unaffected.

Speed is essential throughout the fledglings' journey to prevent them suffering undue stress. After a check up at a holding pen, the young birds are put in boxes and loaded on to a waiting RAF Nimrod to be flown to Scotland. On their arrival at Kinloss there is a quick inspection, and then they are driven to Mallaig, a fishing port on Scotland's rugged west coast. There a boat is waiting to take them to their final destination, the island nature reserve of Rhum, 10,750 ha (26,400 acres) of mountain, cliff and moorland, the home of red deer, rabbits, and over 100,000 breeding pairs of Manx shearwaters – ideal country for the white-tailed sea eagle. By the time the first eagle settles down in its new quarters, a spacious cage on a Scottish hillside, only 12 hours have elapsed since it left Norway.

New arrivals spend about 12 weeks in cages or outside, kept on a running line, where they are able to test their wings for short distances. When the birds are ready for release, each bird is hooded to prevent it becoming alarmed, and the line is detached. It is taken to a nearby vantage point and the hood removed. With a brief pause the young bird rises into the air with sweeps of its powerful wings and is gone, returning for food provided for up to six months while it adapts to the wild.

Since 1975, 42 sea eagles have been released on Rhum, and the reintroduction project is reaching a climax now that the oldest birds are approaching breeding age. There is every sign that, despite their brief ordeal in captivity, the sea eagles recognize their own kind and interact with them normally. Birds from previous years often settle on the cages to inspect the latest arrivals, perhaps forming bonds that will help the younger birds through their first months of freedom. Rhum's sea eagles are kept under the constant watch of John Love of the Nature Conservancy staff, eager for evidence that breeding is beginning. Already birds have been seen displaying – grappling each other's talons in mid-air, or carrying sticks as a prelude to nesting. As the oldest birds reach maturity, it will soon be known if the white-tailed sea eagle is in Scotland to stay.

L.G.

ABOVE *Sea eagle arriving at RAF Kinloss*

OPPOSITE ABOVE *A new arrival on a running line*

RIGHT *John Love with a tethered sea eagle*

5. THE PRIVATE WORLD OF THE WILD

A MOUSE'S TALE   producer Dilys Breese

Wherever you live, in town or country, in thatched cottage or high-rise flat, the chances are that your every move is watched or overheard by dozens of tiny eyes and ears – the eyes and ears of mice. We may not be aware of it, but this unseen audience is truly vast: even in these overcrowded islands, there are more of just one species – the wood mouse – than there are human beings.

Why, then, are mice so numerous, and why do they live in such close conjunction with ourselves? Many animals share our homes and surroundings by invitation: they guard our property, provide food and companionship and in return we feed and shelter them. Unintentionally, we do the same for mice. Many of our dwellings, even the most modern and streamlined, enable house mice to live in far better conditions than they could do in the wild; outdoors, our man-managed countryside offers, in gardens and hedges, in crops and woodlands, habitats that otherwise would not exist.

In 'The Mouse's Tale' we wanted to take a look at a few of these mice, to see just how their life styles interlocked with those of man. We chose a place where several sorts of mice found their 'ideal homes' in close proximity – in and around a farmworker's cottage on a big farming estate in Wiltshire. The cornfield in front of the cottage provided the traditional habitat for the harvest mice, while wood mice (also called long-tailed field mice) had

their network of nests and runs on the edge of the garden. Dormice – not strictly speaking mice, in fact – need a very special place to live, and they found exactly what they wanted in the hazel coppice behind the cottage. And the house mice? No prizes for guessing where they lived.

The human hosts to this horde of four-footed guests were a young dairyman and his wife, Mr and Mrs Glasscock and their baby son Jarno, who was born just before we started the film. With them, in and around the cottage, lived a Noah's ark of domestic animals; a big Alsatian (German shepherd) dog, four goats, a handsome cockerel with several hens, plus eight (I think) cats and kittens, who took a keen professional interest in the mouse filming.

There were many more human beings behind the cameras, of course: in particular wildlife cameraman Owen Newman, who was responsible for nearly all the purely mouse sequences in the film. Many people have asked if the cottage shown in the film was a real cottage, and if the mice were truly filmed there. Yes, indeed it was, and they were. For many days, and sometimes weeks, at a time, Mr and Mrs Glasscock had their living room invaded by house mice performing bits of typical mouse behaviour. For example, a group of young mice were filmed exploring the unfamiliar room for the first time, checking nervously around the skirting board, inspecting toys and table legs, and finally dashing back to their hole – at all times under the camera's watchful eye.

OPPOSITE *Harvest mouse*

But some of the most intimate parts of the mouse's lives could only be filmed in the quiet of sets, specially constructed for the purpose in Owen's studio. In such a set, we watched a dormouse struggling to emerge from hibernation, a painful and hazardous experience in which it was impossible not to feel involved. In another, a female dormouse produced her litter of four pink sugar-like babies: the first time, we believe, that a dormouse birth had ever been filmed.

Although a fairly small number of individuals appeared on the screen, there was a sizeable mouse repertory company in reserve. Not surprisingly, the house mice were the most prolific; indeed, they reproduced so rapidly that 50 had to be released each fortnight. There were about 30 dormice, mostly bred in captivity, but only a small number of harvest mice. In all, something like 300 mice took part.

The art of mouse filming – as indeed of wildlife filming generally – is to persuade the animal to do something that it does quite naturally in the normal course of its life, but in a situation that enables you to film it. Sometimes you may have a bit of luck (and, heaven knows, you cannot manage without it) but mostly it is a mixture of patience and careful contrivance. For instance, Patsy, the handsome black-and-white cat who was seen in the film pursuing a mouse into its hole, became disenchanted after half a dozen 'takes', and to obtain the final shot of Patsy putting its paw through the mousehole, Jen, Owen's wife, had to secrete herself in the cupboard behind, making exciting rustling noises with silver paper. Incidentally, no mouse was killed during the making of the film, though inevitably, in the course of the 18 months, a number died of natural causes: for these sorts of mice one year is a good age.

So much for the filming stories, but what about the real-life mouse stories which we were trying to show on the screen? How do they take advantage of man, and how does man affect their lives, for better or for worse?

To take the 'country mice' first, the harvest mouse is Britain's smallest rodent, a delightful little russet-coloured animal weighing only 6 g (¹/₅oz) as a full-grown adult. It is traditionally associated with cornfields, and indeed the harvest mouse is superbly adapted to these man-made miniature jungles. Its skeleton is light, only 5 per

cent of the total bodyweight, and it has a prehensile tail which it uses very much like a fifth limb. It curls the tip tightly round grass stalks to anchor itself while feeding or uses it as a brake while it scampers lightly to and fro. The 'red mouse', as it is sometimes called, has an excellent sense of balance and uncanny skill in estimating if a cornstalk will bear its weight.

A cornfield supplies virtually all the harvest mouse's needs. As well as taking grain from cereal heads, it also eats a wide range of insects, fruits, seeds and berries, all of which are found in and around the field. Cornstalks are also an ideal site for the harvest mouse to build its amazing nest. It is about the size of a cricket ball, and the fact that it is woven out of the leaves of the still-living plant means that the nest will provide the young mice with a home which is not only comfortable, but also well camouflaged. The Reverend Gilbert White, the eighteenth-century clergyman naturalist, was the first person in Britain to describe the harvest mouse as a separate species. A nest that he examined was, he says: 'So compact and well filled that it would roll across the table without being discomposed though it contained eight young.'

But by no means all harvest mice live in cornfields. They can also be found in reedbeds and rushes, grassy hedgerows and brambles – and indeed, the easy life in twentieth-century cornfields has one considerable disadvantage: the combine harvester. As their habitat is destroyed within a matter of hours, adult mice and independent

OPPOSITE *Harvest mouse in winter coat*

RIGHT *It's amazing nest*

youngsters can scamper away to winter in the rough ground surrounding the field, but litters in the nest will be instantly destroyed. Man-made situations often present habitats that are temporarily ideal for mice, but they carry the penalty that they are always subject to disturbance or even total destruction.

Another 'mouse' that has found man to be both its friend and its enemy, through his management of the countryside, is the dormouse. The 'dozing mouse', the 'seven sleeper', was regarded with affection by our forefathers, and seems to have been much more common in the past than it is today. In Victorian times the dormouse was a popular schoolboy's pet, though today most children have only made its acquaintance through the pages of *Alice's Adventures in Wonderland*.

The dormouse is a delightful little animal, with bright orange-brown fur, large black eyes, long whiskers and a furry tail, which distinguishes it from any other small British mammal. Unfortunately the dormouse's nocturnal life style means it is seldom seen in the wild, but its presence can be detected by a number of clues, such as the shells of hazelnuts which it opens in a distinctive way. Holding the nut in its front paws, the dormouse gradually turns it round, cutting through the shell with its lower teeth, and removes the kernel bit by bit: often as long as five minutes will be devoted to a single nut. The discarded shell, with its smoothly chiselled edge, is a real dormouse trademark.

Another sign of the unseen dormouse is its nest — or rather, one of its nests, because during the course of its life it will build three different types for different purposes. The first of these, sited at or near ground level, perhaps among tree roots or in an old hollow stump, is its hibernation nest. The dormouse is thought of as a symbol of cosiness because it spends five months of the year 'asleep'; but although hibernation is an effective way of passing a period when weather is unfriendly and food in short supply, it is also an ordeal. During hibernation, the dormouse's whole metabolism slows down, its respiration and heart rate fall, its temperature drops, even the composition of its blood is altered. The return to normal takes up to 12 hours and imposes a great strain on the little animal. Altogether, from one cause or another, up to 80 per cent of dormice are thought to die during hibernation.

The next nest is of a quite different kind: the nest the pregnant female makes to give birth to her young. She usually sites it about one metre off the ground in a bramble bush or the fork of a sapling. This nest is large and rather untidy looking, often made of honeysuckle bark and dry grass, with a soft, finely shredded lining. The dormouse babies, usually a litter of four, look very undeveloped at birth. Their eyes are closed, they only have folds of skin where their ears will be, and their toes are all stuck together. It will be 18 days before their eyes open and a month before they will leave the nest.

When the young make their first excursions from the nest, in search of their favourite blackberries and nuts, they already have a good sense of balance. Dormice are well suited to life in the trees; both front and back feet are adapted to grasping twigs and have small pads to help them keep a firm grip. They scamper confidently about the complex network of twigs and branches, in what appears to human vision to be total darkness, aided by their

large eyes and long whiskers. The fine bushy tail is not used for grasping, as with harvest mice, but for balancing. As the young start to become independent, the third sort of nest is made: a flimsy 'bachelor pad' in which the dormouse lies up during the hours of daylight.

The dormouse's life style, then, is a specialized one. Its requirements – principally hazel bushes and brambles for food and nests – mean that its natural home is the edge of woodland, a habitat that is no longer common in Britain. For hundreds of years man provided this habitat artificially by coppicing, an ancient rotational method of cropping hazel and sweet chestnut trees to provide fencing, jumps for horses, and thatching spars. But nowadays these items are no longer so much in demand and the coppiced woodlands the dormouse needs are in short supply.

But the ideal home for another sort of mouse is easy to find. The ancestors of the house mouse are thought to have evolved on the steppes of central

ABOVE *Dormouse opening an hazel nut*

OPPOSITE *Well suited to life in the trees*

ABOVE RIGHT *Dormouse in hibernation*

RIGHT *Dormouse nest*

151

Asia, from where it hitched a lift with man and is now probably the most widely distributed of mammals, found almost all over the world. House mice now live virtually everywhere that man has set his mark. They thrive in the concrete labyrinths of high-rise flats, in the depths of coal mines, and in cold stores where the temperature is well below freezing.

The shelter and regular food supply to be found in building mean that house-living house mice, unlike those living in the wild, can breed throughout the year. Straw, paper, electrical insulation form the nest: mice have even been known to use

pound notes or top-secret blueprints as bedding. And the potential population is amazing. If a pair breed at the start of the year and their young survive and breed, and so on, then by New Year's Eve, there would be a grand total of 2500 mice!

Of course, they do not all survive: many mice are destroyed by predators. We have at least two misconceptions about house mice, and one of them is that they are efficiently controlled by cats. An ancient Welsh law records the prices to be paid for cats: one penny for a kitten, twopence for an inexperienced youngster, but fourpence for a cat after it had caught its first mouse. The truth is that even a proper 'fourpenny cat' is not particularly efficient at clearing a building of mice. The natural predators of mice include the barn owl, weasels and stoats, and (oddly enough) rats, but man with his artificial methods is by far the most efficient control.

And the second of these misconceptions about house mice? That their favourite food is cheese.

House mice certainly do eat cheese when it is available, but then they will eat virtually everything that human beings eat, plus odd items like plaster, soap, glue and candles. A mouse needs about a fifth of its bodyweight in food each day and, remarkably, hardly any water, so it is not surprising that a colony of mice can live comfortably in blocks of offices, for example, off the remains of lunchtime sandwiches.

Of all the mice that live with us, house mice are undoubtedly the champion opportunists. Whatever new foods, new breeding sites, new possibilities of any kind human beings offer, the house mice hordes are always poised ready to take advantage of them. Their huge success as animals, their spread throughout the world, are the direct result of making use of conditions provided by man, following him wherever he goes. Mice have already accompanied astronauts on their lunar voyages: will they one day join them in colonizing the moon?

ABOVE *House mice young*

OPPOSITE ABOVE *House mouse eating peanuts*

OPPOSITE BELOW *The potential population is amazing*

RIGHT *Eats about a fifth of its body weight each day*

NIGHT IN THE UK COUNTRYSIDE *producer Dilys Breese*

The **water shrew** is well named. Sometimes, as it hunts for small fish, insect larvae, amphibians and snails, it can be seen swimming on the surface of its pond or stream; sometimes it dives down and 'walks' along the bottom, its coat silvery with the many air bubbles trapped in its fur.

The water shrew is active both by day and by night, for it must feed every 2 to 3 hours: in order to survive it has to eat its own weight in food every 24 hours. Each time it emerges from the water it has to dry its fur so that it does not lose body heat and so die of cold. Some of the water is squeezed out as the shrew passes through its network of narrow burrows; the remainder is combed out with its hind feet. So great are the pressures on the water shrew that it does not feed first and then groom, but economizes on time by doing both together.

The **fox** is one of the real survivors of our modern world, now found in all sorts of habitats from city centres to deep countryside. Although quite often seen during the day, foxes are chiefly creatures of the night. Their sharp barks and bloodcurdling screams are most commonly heard in the three hours after sunset, especially during the winter breeding season.

The fox is both a skilful hunter and a versatile scavenger, which provides him with a very varied menu. Some individuals specialize in poultry and lambs, but more generally foxes take rabbits, hedgehogs and birds, snails and beetles, carrion and even fruit.

The **tawny owl** is the commonest species of owl on the mainland of Britain, found in habitats ranging from deciduous woodlands to heavily built-up city areas. Although its large eyes are clearly adapted to night vision, the tawny owl hunts mainly by hearing and then seizing the prey in its powerful talons. In woodland they chiefly take small mammals such as wood mice (seen here), but owls in cities concentrate on house sparrows and brown rats. The owl's night-time presence is usually made obvious by its call, generally – but not very accurately – written as 'toowhit, toowhoo'.

It is tempting to describe the slug as 'a snail without a shell', although in fact many slugs have a vestigial shell, usually hidden inside the body. In the case of roundback slugs like the **great black slug** – its colour is very variable, in fact – the shell is reduced to a number of chalky granules.

The absence of the external shell makes slugs vulnerable to dessication and to predation by many animals such as hedgehogs, frogs, toads, and many sorts of birds. One of the ways in which slugs respond to these hazards is by feeding largely by night, or by day only if it is wet. They eat a wide range of foods, including green and decaying plant matter, tubers, carrion and dung, attracted to them by the scent organs in their tentacles.

The name '**glow worm**' (*left*) must originally have been inspired by the wingless larva-like female, but glow worms are actually beetles. At nightfall the female climbs a prominent piece of vegetation and raises her tail end, displaying the luminescent patches on the last three segments to attract the male to fly in and mate with her.

Adult glow worms eat little, if at all, but the larvae feed on snails. They inject their prey with a fluid which serves partly to paralyse and partly to digest it: the larvae then suck up this 'snail soup'. Glow worms are, in turn, eaten by night-feeding animals like toads and hedgehogs.

One possible cause of the recent decline in glow worms is that the male is led astray by artificial lights, which divert him from the female's more modest glow.

With its large eyes and ears the **wood mouse** – also called the long-tailed field mouse – looks like a more elegant version of the house mouse, and is also sometimes to be found in houses. More typically wood mice live in woods, fields and gardens, making their colonial homes in a complex system of underground tunnels and runways. They are preyed on by many animals such as stoats, weasels, foxes and domestic cats, but owls – and especially tawny owls – are their chief predator. Indeed, so important to the tawny owl is this source of food that when mice are in short supply the number of tawnies breeding declines.

The **barn owl** has two claims to fame: as the most widely spread land bird in the world and as a champion night-time hunter. Although hidden under its feathers, the barn owl's ears are very well developed and placed asymmetrically, enabling it to judge the position of prey accurately by hearing alone: indeed, barn owls have demonstrated their ability to catch their prey in total darkness.

The value of this acute hearing is increased by the fact that the barn owl itself is almost totally silent in flight, as a result of the fine down that fringes its powerful flight feathers. Often nesting in churches and old buildings, many a ghost story must have originated with the barn owl.

When it is alert and active the **long-eared bat's** most prominent feature is indeed the huge ears, which are three-quarters the combined length of the head and body, but when asleep the ears are neatly folded down under its wings.

Like other bats, the long-eared is a highly sophisticated nocturnal flying machine, but in addition to using its powers of echolocation to catch insects on the wing the long-eared bat more commonly flies to and fro through foliage, hovering to pick insects off the leaves. Small insects are eaten in flight, but larger ones like moths are 'pouched' in a skin bag between the legs until the bat returns to a perch.

Hawk moths are spectacular creatures and among the fastest fliers in the insect world. Almost all of them are night-flying, and are best seen as they hover like hummingbirds in front of an array of

flowers. The moth uses its long tubular proboscis, which is coiled like a watch spring when not in use, to probe deep into the flower and feed on its nectar. However, it takes a while to get one of these big insects 'ready for take-off': after a period of rest a hawk moth has to vibrate its wings for a minute or more to raise its body temperature to the level at which it is able to fly.

Many hawk moths, like this privet hawk moth, are named after their food plants – that is, not the flowers from which the adult takes nectar, but the leaves that provide food for the equally extraordinary caterpillar.

Birds in general sleep, or roost, at times when they do not have more important things to do. Consequently they *usually* sleep at night, but owls, for example, which feed at night, sleep during the day. Birds like waders and wildfowl, whose food availability is affected by tides, sleep during those periods when it is impossible to feed.

Having settled to rest, the bird relaxes into a comfortable position and fluffs up its plumage to provide better insulation. A bird does not really 'tuck its head under its wing'. Some birds rest their heads along their backs or, like this **mute swan,** turn right round and tuck their bill into their shoulder feathers.

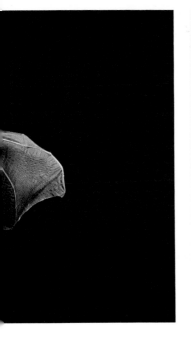

THE GREAT HEDGEHOG MYSTERY *producer Dilys Breese*

In July 1979 Mr Ping Law, the proprietor of a Chinese restaurant in Sunderland in northeast England, had a frightening experience. Going into his kitchen one evening, he found inside a very peculiar and, he thought, highly dangerous animal. He sent for the police. Great was their surprise to find that the intruder was not a man-eating tiger, not a black widow spider, nor yet several yards of python, but – a hedgehog!

The hedgehog certainly is an odd-looking beast, dumpy, short-sighted, and above all covered with that strange tea cosy of spines, but so familiar that most British people have come to accept it and even to regard it with affection. Indeed, despite its obvious disadvantages – smelly, nocturnal, flea-ridden and the very reverse of cuddly – the hedgehog must be a strong candidate for the title of most popular wild mammal in Britain.

Certainly, no animal can have intrigued and mystified its human admirers like the hedgehog. Over the centuries whole bookfuls of legends and superstitions have accumulated about it, and what's more, new legends are still growing up today. So in 'The Great Hedgehog Mystery' we decided to investigate some of these beliefs, both ancient and modern, to see how much truth there is in them, and how they arose in the first place. To do this we enlisted the assistance of hedgehog enthusiast Pat Morris and a small repertory company of hedgehogs. Very early in the filming we discovered that the average hedgehog's reaction to light – and also to high-pitched sounds – is to curl up and wait for it all to go away, so we had to assemble a small group of animals that could be accustomed gradually to a certain amount of light and human bustle. One or two of these were 'pet' hedgehogs, though all were born in the wild: Emily was a particularly pretty animal, though small – the rescued runt of a litter; Georgie, on the other hand, was a very large lady, with much of the temperament traditionally associated with film stars. And I must not forget the male Starsky. We didn't get round to acquiring a 'Hutch'. Most of the hedgehogs, however, were 'borrowed' from the wild for the duration of the filming. For example, one male was rash enough to knock over my milk bottles very loudly while consuming the nightly bowl of bread and milk, and, along with the others, went to spend a few weeks in the large garden of Owen Newman, our wildlife cameraman.

So, what are these hedgehog legends and mysteries that we wanted to investigate? Probably the most ancient and the most widespread concerns hedgehogs carrying apples on their spines. This story was first recorded by Pliny, writing nearly 2000 years ago, and has been repeated over and over again, acquiring further detail over the centuries, up to the present day. The hedgehog, it is said, rolls on fallen apples – sometimes grapes, figs or other fruit – so as to impale them on its spines. It then carries the apples back to its nest to serve as provisions for the winter or to feed its young. As a seventeenth-century writer says: 'When he findeth apples or grapes on the earth, he rowleth himself upon them until he have filled all his prickles, and then carrieth them home to his den, never bearing above one in his mouth . . . And if he have any young ones in his nest, they pull off his load wherewithal he is loaded, eating thereof what they please and laying uppe the residue for the time to come.'

This raises two questions: first, is a hedgehog physically capable of carrying fruit on its spines, and secondly, if it is, would it actually do so? We put it to the test.

The scene of our experiment was a small orchard near Guildford in Surrey. It was autumn and the ground was covered with slightly overripe apples. It would, of course, have invalidated the experiment if we had tried to persuade the hedgehogs to roll on apples and impale them on their spines unless they showed a natural tendency to do so – which they did not. What we wanted to establish was whether, if they collected fruit on their spines

OPPOSITE '*Hedgehogs have been around for 15 million years*'

RIGHT *Detail from a medieval bestiary*

by design or by accident, they could then hold the fruit in place and walk freely carrying it on their backs. The answer seems to be yes. Several different hedgehogs demonstrated clearly that, given a suitably squashy apple, they could walk quite comfortably and at a good speed, carrying the apple on their spines. This is not really surprising: the muscles associated with the hedgehog's spines are very powerful. It is these muscles that enable the hedgehog to curl up into a tight ball, and so remain almost impregnable in the face of possible predators. A couple of apples – provided they are not of a size to win the village produce show – can be carried with little difficulty.

But why should they want to? Hedgehogs may eat fruit occasionally, but it is certainly not a major part of the diet: slugs, earthworms and caterpillars are the sort of food they prefer. And as for taking food back to the nest, why should they bother? A hedgehog does not need to collect food for the winter because it hibernates: it has already laid up its winter supplies under its skin in the form of fat,

so stores of apples in the nest would be irrelevant. Sadly, then, it seems that the image of the apple-carrying hedgehog, familiar for hundreds of years, is just a nice story.

Another belief about hedgehogs, which is very widespread in country lore, is that they suck milk from cows' udders. Here again, we had the same two questions to ask: *could* hedgehogs do this, and is there any reason why they should? In this case, the second question was easily answered. Anyone who feeds a garden hedgehog knows that a dish of milk – or more usually his bread and milk – is one of the easiest ways of attracting a hedgehog to the doorstep. But even so, would it be possible for a hedgehog to suck a cow's udder? Assuming that the cow were lying down, could the hedgehog's mouth open wide enough to grasp the teat? Experiments have been carried out to show that hedgehogs certainly can suck the sort of rubber teats used to feed orphan lambs, and cases of cows with udder injuries thought to be due to hedgehog teeth have been recorded in the *Veterinary Review*.

OPPOSITE *Hedgehog hibernating*

BELOW *Swimming*

But would a cow actually sit and play the part of an animated milk bar?

The 'star' chosen to take part in this particular experiment was a totally wild animal called Elsie, a rather plain but obliging beast, who seemed happier than the rest of us to be wandering round the fields at 4 o'clock in the morning. The herd of black-and-white Friesians were dozing quietly, some chewing the cud, many sitting on the ground. Left to her own devices, Elsie trundled to and fro among them, not at all nervous; indeed some of the cows were much more conscious of the hedgehog than she of them. After a while she headed towards one particular cow whose distended udder, shortly before milking, had leaked some drops of milk on to the grass. She sniffed and moved in – but not to the udder, to the milk, which she lapped up with typical enthusiasm. Later the herdsman in charge of these cows told us that he had several times before seen hedgehogs behaving in just this way on the same pastures where we were now standing, and it seems probable that,

whether or not hedgehogs really do suck cows' udders, this is by far the more usual way for them to get a free drink of milk.

But perhaps the single aspect of hedgehog behaviour that attracted most attention from ancient authors was mating. Finding it impossible to imagine that such prickly animals could copulate in the manner of dogs or other mammals, they conjectured that hedgehogs mated face to face. 'When they are in carnal copulation,' wrote Topsall in 1658, 'they stand upright and are not joined like other beasts for they embrace one another standing belly to belly, for the prickly thorns upon their backs will not suffer them to have copulation like dogs or swine.'

The mystery of hedgehog mating still continues to fascinate, hence the old gag: 'How do hedgehogs mate?' Answer: 'With great caution.' Even today the full procedure of courtship and mating has seldom been observed and, we believe, never before filmed. In the course of filming for 'The Great Hedgehog Mystery', Owen Newman was able to record several hedgehog matings which are of great scientific, as well as popular interest.

Many attempts at mating in hedgehogs come to nothing: if the female is not receptive she simply runs from the male, and it is surprising how fast hedgehogs can run when they want to – about as fast as a brisk walk in a human being. But even when the female is receptive 'courtship' is a long and noisy procedure. The male circles round and round her close to her body, while she keeps turning towards him flinching to indicate her unwillingness; the whole business is accompanied by a great deal of grunting and snorting. It has been claimed that a particular male will only circle in one direction, either clockwise or anti-clockwise, but we saw that this was not the case: the courting male will turn and circle around the other way. This performance may continue, in bouts of 20 minutes or so, for several hours. Finally, the female indicates her readiness to mate by arching her back, spreading out her hind legs, and flattening her spines. The male mounts, grabbing the female by the scruff of the neck and they copulate. After mating, the two animals separate, and very probably will never meet again: the male plays no part in bringing up the family he has sired. So hedgehog mating is nothing like the legend: you might say that it is even more remarkable.

But if it is hard to imagine mating between two spiny animals, perhaps it is even harder to imagine birth. Our forefathers do not seem to have worked out an alternative technique for producing young as they did for copulation, but there certainly is a legend that giving birth is extremely painful for the mother, and that she actually pricks herself with her spines to distract herself from the pain of producing her equally spiny young. As it happens, it seems that giving birth is indeed particularly painful for hedgehogs as we saw from the prolonged heaving and straining of our particular mums, but not because of their babies' spines. At birth the babies are almost smooth: the only indication of spines is a pattern of little bumps, rather like goose pimples, in two bands along their backs. These are the sites from which, only a couple of hours after birth, the young hedgehog's first set of soft white spines will start to emerge. At this stage there are only 90 of them, but the babies grow rapidly, soon adding further sets of brown adult-type spines till, at three weeks or so, they look like miniature versions of their mother. But at birth they are totally unprickly and the legends about hedgehog birth seem to be the result of sympathetic conjecture by some bygone sage.

Indeed, throughout the centuries people have always shown an affectionate interest in hedgehogs, combined with a lively curiosity about their lifestyle. Our 'rude forefathers' obviously expected such an odd-looking animal to behave oddly, too, and to carry out its most ordinary functions in an extraordinary way. And this attitude still persists today. Not only do many of the ancient hedgehog myths linger on, but new ones continue to arise and flourish. A good example concerns hedgehog fleas.

The fleas themselves are far from legendary. Parasites are the penalty the hedgehog pays for its otherwise highly efficient suit of armour, and five hundred fleas have been removed from a single animal. The belief has recently grown up that hedgehogs actually *need* their fleas and cannot survive without them: not so. Sometimes well-meaning people spray infested hedgehogs with an insecticide which is injurious to them, and so kill off both host and parasites together: perhaps this is how the legend arose.

We did not need to set up an experiment to check out this legend, however, because it had already been tested, in quite an unusual way: towards the end of the nineteenth century, British hedgehogs were exported to New Zealand by a party of homesick immigrants, and during the long sea voyage they lost their fleas. However, this certainly did not prevent the hedgehogs from settling down happily in their antipodean homes, and it seems that the hedgehog population is even denser in New Zealand now than in Great Britain. Incidentally, human beings need not worry about hedgehog

LEFT *Five-day-old hedgehog*

OPPOSITE *Hedgehog mother and young*

fleas taking up residence on themselves or on their pets: hedgehog fleas much prefer to remain on hedgehogs, and do them little harm.

Another very weird 'legend' has only come to light this century, and it is called 'self-anointing'. This 'legend', however, is undoubtedly true. A hedgehog, attracted by some special substance such as leather, cigar butts, or even distilled water, will lick it vigorously until it salivates and produces quantities of foam. It then rolls over and spreads the foam as far over its spines as it can manage, often indulging in the most extraordinary contortions to do so. Some hedgehogs seem to be self-anointing specialists, and two of our 'stars'

ABOVE *Tick on hedgehog*

RIGHT *Self-anointing*

OPPOSITE *Albino hedgehog*

seemed to be particularly enthusiastic. Emily was 'turned on' by carpet underfelt – rather difficult for filming! – also by leather. She was fascinated by our cameraman's sandals, and even climbed inside one when he took it off. Georgie was encouraged to self-anoint by a variety of substances, again including leather, and the performance we actually filmed was inspired by a creosoted fence.

So why do they do it? Many theories have been put forward: for example, that it is some sort of sexual display, but this does not seem to be the case. Another theory is that hedgehogs self-anoint after chewing the skins of toads, using the toad venom to increase the protective value of their spines. Or again, that they are using the saliva as a form of insecticide to control the numerous pests in their spines. If so, it does not seem very effective: as so often with hedgehogs, it is still a mystery.

Perhaps the most widespread and popular of modern hedgehog legends concerns hedgehogs and traffic. The pathetic sight of one of these harmless little animals lying squashed on a busy road distresses most of us, and is all too common. It has been suggested, over the last few years, that evolution is at work before our very eyes, and the hedgehogs are learning not to roll up in front of

164

approaching traffic but to run and so escape. It is a cheering thought, but probably an example of wishful thinking. The hedgehog is a very ancient type of animal. They were trundling about this earth fifteen million years ago, long before the mammoth and the sabre-toothed tiger. In comparison, the history of the internal combustion engine has lasted only the twinkling of an eye. And one soon realizes that the hedgehog that escapes from one set of wheels may well run straight under another. So it is hard to place much confidence in this legend, encouraging as it is. Is it true that hedgehog populations are being affected by deaths on roadways? We do not know, but probably not.

The fact that we see many dead hedgehogs on the roads means, after all, that there are almost certainly many more, alive and well, thriving in spinneys and woodlands and suburban gardens throughout Britain.

We should not forget that the hedgehog's traditional armour is highly effective against conventional enemies. Apart from man-made cars, it has few predators, and deals with those pretty effectively. So this homely little animal, having been around for fifteen million years, looks as if it may well survive for another fifteen million, producing more and more mysteries for us to investigate — let's hope so!

THE DRAGON AND THE DAMSEL *producer Pelham Aldrich-Blake*

Winged dragons still roam the British countryside – diminutive in size compared to the beasts of myth and fable, but every bit as ferocious in their small way. The dragonflies, and their cousins the damselflies, are the subject of a 'Wildlife on One' programme to be filmed during 1982.

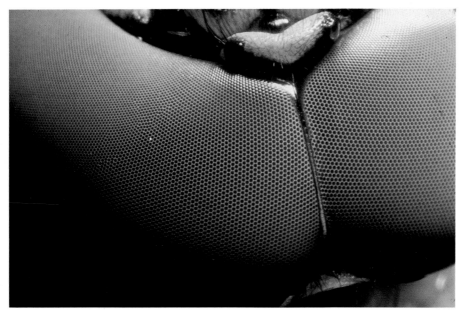

LEFT A typical dragonfly, the brown aeshna. Dragonflies possess great aerial agility and catch their prey – *gnats and other small insects* – on *the wing. The hairy legs are held pointing forwards to form a 'net' in which to seize them*

BELOW *The compound eyes of a dragonfly in close-up. Vision is the dragonflies' dominant sense: they hunt and find their mates by sight, and can detect movement up to 12 m (40 ft)*

OPPOSITE ABOVE *The gold-ringed dragonfly, one of the larger and more spectacular British species, found along streams on heath and moorland*

OPPOSITE BELOW LEFT *A damselfly, the demoiselle. Damselflies are of more slender build than dragonflies, and at rest their wings are held back along the body rather than out sideways*

OPPOSITE BELOW RIGHT *Although slower flying than dragonflies, damselflies too catch their prey on the wing. Note the large eyes and bristly legs of this demoiselle*

RIGHT *Damselflies and dragonflies assume strikingly unusual postures during mating. Most insects mate 'tail to tail', but the male dragonfly possesses secondary genitalia beneath his thorax. He swings his abdomen forward to touch them and fills them with sperm, and the pair then assume a characteristic 'wheel position' during copulation with the female's tail against the male's thorax. In many species mating takes place on the wing, and male and female often remain together in tandem during egg-laying*

BELOW LEFT *Despite their prowess on the wing, dragonflies and damselflies sometimes fall prey to other animals, or even plants.*

BELOW *Some species scatter their eggs loose into the water, others place them with greater precision on floating or submerged vegetation. A few cut into plant stems with their ovipositor and insert their eggs within the plant's tissues, perhaps even descending beneath the water to do so*

ABOVE and LEFT *The larvae of dragonflies (above) and damselflies (left) are aquatic. Some are short and squat and conceal themselves in mud or decaying vegetation on the bottom; others are slender or streamlined and live among weeds in mid-winter. All are voracious predators and will tackle almost anything that moves, up to their own size. The larval stage may last for one to three years, in contrast to the few weeks of adult life*

169

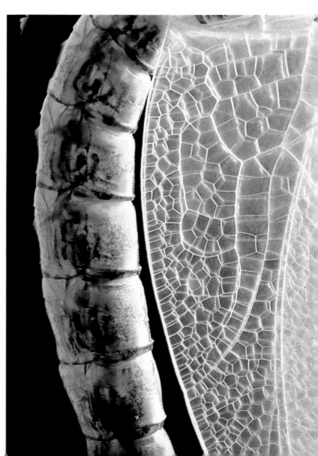

OPPOSITE ABOVE LEFT *A dragonfly leaves the water and climbs a weed stem. For the past four days, the tissues within its hard outer skin have been changing to the adult form*

OPPOSITE ABOVE RIGHT *The cuticle splits and the adult begins to emerge*

(**frontispiece**) *With its head and thorax out, the dragonfly hangs upside down from its abdomen and rests for up to an hour*

LEFT *Free of the larval skin, the newly hatched adult rights itself and waits for its wings to expand and harden. At this stage it is vulnerable; hence emergence usually takes place at night*

OPPOSITE BELOW LEFT and RIGHT *The wings enlarge. Blood is pumped into them through the network of veins, and within half an hour they will be fully expanded*

BELOW LEFT *The wings are nearing their final size, but they are still soft, and the adult still clings to the old larval skin*

BELOW RIGHT *The complete insect crawls up the reed stem for its maiden flight. Although its wings and cuticle are now hard, it will take several days to reach sexual maturity and attain the full brilliance of adult colouring*

AMBUSH AT MASAI MARA   *producer Keenan Smart*

It was dark over Europe, dark over North Africa. The Kenya Airways Boeing 707 was half empty as it cruised through the night sky. Through the scratched window glass beside me I had glimpsed the bright lights of Rome an hour earlier, but since then only the occasional twinkling star had broken the steady darkness.

It felt strange lying stretched out on the big jet floating over the ruins of ancient civilizations 10,700 m (35,000 ft) below. The Pyramids were down there somewhere as we flew on, so was the mighty Nile winding its way through the dryness of the Nubian desert. Its two lesser relatives flowed there also, the Blue Nile pouring down from the Abyssinian Highlands and the White Nile draining from the shores of Lake Victoria on the equator only a few hundred miles from where I would soon be camping.

According to my watch the hours were advancing but it seemed I was travelling ever backwards in time to Africa – the cradle of mankind. Back in time to the last great reminder of what the world was like when man's ancestors first pitted themselves against the primeval wilderness, stumbling and struggling in an environment fraught with dangers.

Beneath, in Kenya, was a world still filled with enormous herds of animals and that glorious powerful golden cat: *Panthero leo*, the lion of Africa. King of beasts, symbol of the sun, he was the reason for my journey.

A few days before I had stood in the draughty cold and pelting rain at Bristol Zoo gazing in wonder at the captive lions. A line from the famous zoologist Konrad Lorenz crept into my mind: 'There are certain things in nature in which beauty and utility, artistic and technical perfection combine in some incomprehensible way, the web of a spider, the wing of a dragonfly, the superbly streamlined body of the warhorse, and the movements of a cat.' But Lorenz may have been thinking of a stalking lioness.

On that miserable morning, it was difficult to reconcile those words with the sight in front of my eyes. I was confronted with a caged and shackled legend. Here was primitive man's worst fear. The fiercest and most lethal of predators; the very embodiment of power and terror imprisoned in a concrete jungle. Yet even here these animals were impressive. Their stares held a noble, haunting quantity that pierced the distance between us and thudded straight into the oldest recesses of my brain. The hair on my neck stirred – I had seen that look before and I knew how early man must have felt, alone and afraid in the wild.

Deep in our innermost being lurks respect and fear for the lion, those ravening, roaring beasts of the Bible. The lion's presence demands attention, his authority matches our own. Man and lion – the one wary of sheer physical power, the other cautious in the presence of a dominant intellect. Since man first went to war, might has been symbolized in the lion's shape. The lion rampant adorns banners, decorates coats of arms, a source of inspiration and courage.

The zoo lions were still proud and defiant in their confinement but that fierce strength they possessed was locked inside the steel cell that surrounded them. I wondered what it would be like to see free lions unfettered by the bars and cages of civilization living as they have always done in their natural domain. How would I react to being this close to a wild lion with nothing between us but the thin canvas walls of the tents in which we would be living and sleeping right in the heart of their territory in Kenya's Masai Mara game reserve? I quaked at the thought. I was on my way.

The long weeks of preparation and planning were over. Nothing more could be done to smooth the way for the arrival of the cameraman, Hugh Miles, and his mountain of equipment. There would only be two of us on this trip to begin with – Hugh and myself – and we had an all too brief three weeks to complete our filming.

Hugh is a brilliant cameraman whose skills are much in demand. His busy timetable left little room for manoeuvre. A few weeks before leaving for Kenya he had been sharing ice floes with polar bears in the Canadian Arctic, filming for another BBC series. He had then gone to Shetland in search of otters and at the end of the Kenya week he was

That glorious powerful golden cat

eager to return for the big international wildlife film makers' symposium held in Bath at the end of September. So three weeks it had to be, timed, we hoped, to coincide with the arrival in the Masai Mara of the enormous wildebeest herds.

Everything depended on the wildebeest being there but there was no guarantee that we had got the timing right. No one could predict accurately when the mighty army of migrating animals would reach the Masai Mara, or indeed if they would do so at all. Many factors influenced the movement of the herds, but the two most important – rainfall and the availability of green grass – varied significantly from year to year.

Nothing quite matches the awesome spectacle of East Africa's annual wildebeest migration. Each year hundreds of thousands of animals trek north from their calving grounds on the Serengeti plains in search of green pasture. The Mara is the northernmost extension of the Serengeti ecosystem and is Kenya's most magnificent wilderness. An area of incomparable beauty and richness of wildlife, it probably contains as many lions, cheetah, buffalo and wildebeest as are found in the remainder of Kenya's pales and reserves added together. While much of Africa bears witness to a sorry tale of depletion of animal numbers and human exploitation of the natural environment, the Mara remains relatively unscathed – a refuge of Pleistocene times in the modern world.

When the curtain goes up on the wildebeest migration the Mara provides a boundless stage on which is enacted the Earth's most dramatic wildlife show.

For the wildebeest, the 800 km (500 mile) round journey they undertake from their breeding grounds is packed with dangers. Many fail to complete the epic march. Wherever they go, lion, cheetah, leopard, hyenas and wild dogs harass the endless columns. In their path lie foaming rivers alive with crocodiles. Thousands drown in the swilling waters or are trampled underfoot on the slippery mud of the river banks as the panic-stricken herds try to cross. The Mara river is perhaps the most formidable of these barriers. The survivors of the hazardous river crossings must run the gauntlet of the lion pride occupying the nearby Musiara marsh.

Musiara marsh lies in the heart of the Masai Mara. Fed by a spring that never runs dry it is a

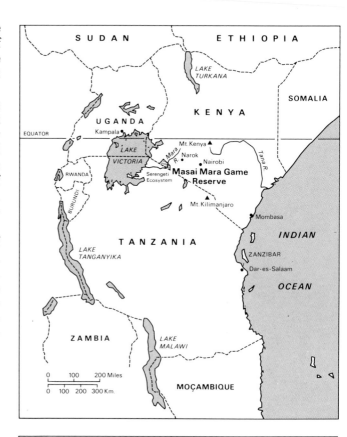

1 Kichwa Tembo Camp	5 Mara Serena Lodge	9 Hippo pool
2 Olololo Gate	6 Ngiro-Are Anti-Poaching Post	10 Mara Research Station
3 Campsite	7 North Mara Bridge	11 Marsh Pride Territory
4 Wildebeest river crossing area	8 South Mara Bridge	

haven for waterbirds and nourishes the diverse life of the surrounding plains. All the plains game must come here to drink. For the marsh lion pride and their cubs it is a prized possession, the most sought-after territory in the Mara – a playground and a hunter's paradise. For the lions the coming of the wildebeest is a time of plenty; gone are the lean spells of the dry season when the best they can do to obtain a meal is to dig warthogs from their burrows. But even in the worst droughts the marsh offered more than the neighbouring plains.

We wanted to tell the story of a day in the life of this one lion pride at the height of the migration. The story of the collision between two of Nature's mightiest forces: majestic lion and migrating wildebeest.

A network of lion prides is scattered over 1800 km² (700 square miles) of the Mara's hunting grounds. But the marsh lions were no ordinary pride. Since 1975 their personalities, disputes and family history have been chronicled in the diaries of Jonathan Scott, a young wildlife artist and photographer, who lives and works in the Mara. His detailed knowledge of the events leading up to the present pride structure gave an intimate side to the story which we hoped to capture on film. Together with Brian Jackman, nature correspondent of *The Sunday Times*, Scott was preparing material for a book on these lions.

I knew the lions would be there somewhere in the vicinity of the marsh. We would just have to find them. But the wildebeest were an entirely different matter. There was no guaranteeing the presence of large herds.

It was still pitch-black as the Boeing began its descent to Nairobi's Jomo Kenyatta airport. I looked again at my watch: 5 a.m. Soon it would be dawn. An African dawn. In the next three weeks I would be up at that hour every morning as Hugh and I worked furiously to get sufficient material to build the story of one long African day from dawn to dusk. Even the taxi ride from the airport into Nairobi was a naturalist's dream. A kaleidoscope of brilliantly coloured birds dazzled the sleep from my eyes. By the time I spotted my first giraffe striding alongside the roadway I was wide awake with excitement, all thought of sleep banished.

I had arrived two days ahead of Hugh. For the next 48 hours I would be tramping back and forth between innumerable government offices getting the final written permissions and licences that would enable us to film without hindrance for, as in most countries, Kenya's bureaucracy is well developed and numerous park wardens and policemen would be demanding to see our official documents over the following weeks.

Hugh's flight arrived several hours late but despite being exhausted he was as keen as I to press on to the Mara. We loaded the battered Landrover I had hired in Nairobi and I wondered what sort of driving conditions we would encounter once the tarmac road petered out at Narok, two hours' on.

I had already returned one hired Landrover which had dreadful oversteer and tyres as bald as eggs. My 'new' vehicle did little more to inspire confidence. Although more positive in its steering I soon discovered it had little else to recommend it. Its doors refused to lock, the fuel and temperature gauges were broken, and ominous noises along with nasty fumes poured from the engine.

In contrast to the bald tyres of my previous vehicle it possessed remoulds of an extraordinary kind. Each tyre looked as if bits of stripped rubber boots had been stuck over the former tread. In the short 20 km (12 miles) drive to the airport to collect Hugh I felt certain they would strip themselves bare. Every few kilometres I stopped to peer at the wheels. Astonishingly, the remoulds showed no sign of parting from the tyre and never did in the course of the thousands of kilometres we were to cover over the roughest roads and boulder-strewn tracks to be found in Kenya. Only once did we run into mechanical trouble with the remarkable old banger, and that was in full pursuit of a cheetah whose performance on the jumbled terrain was far superior to our own. Travelling fast we hit a hidden hyena hole and the drive shaft parted. Luckily we were able to repair it.

Several hours after Nairobi we passed through Narok and from then on we bumped and bruised our way over dusty tracks into the Masai Mara reserve.

The Mara is an area of gently rolling plains bordered by highlands and escarpments on all sides: 'Mara' is the local Masai tribe's word for this sort of landscape. Although almost on the equator it averages an altitude of 1,500 m (5,000 ft) above sea level and so is high enough to avoid the excessive heat of Kenya's lowlands.

Scattered throughout the reserve are patches of

forest and acacia glades. Rocky outcrops push through the grass providing excellent hiding places for leopard and lion, and nightmare driving conditions for anyone trying to follow them.

As darkness began to fall we reached our camp in the riverine forest between Musiara marsh and the Mara river.

Jonathan Scott was sharing our camp and was there to greet us. So were the lions. Almost from the moment we stepped from the Landrover we were aware of their presence. There are few sounds in nature more impressive than a full-throated lion's roar. That night and every night we were treated to a spine-tingling cacophony of sounds

from a variety of animals, but always dominated by the roaring lions. The lions sounded close. Too close for comfort. Beyond the flickering light cast by our camp fire they were prowling around in the blackness. Elephant and buffalo were also wandering among the trees and the dramatic snorts of hippos could be heard from the direction of the river. Baboons screamed from the trees and occasionally loud noises like pistol shots rang out above the hubbub as some animal snapped a fallen branch underfoot. According to Jonathan, the hippos would soon be leaving the water on their nocturnal wanderings in search of grazing around the marsh. Our camp lay close to the path they took from the river, he added. Should they stray from their traditional route or be panicked by our lights they might well come crashing through the tents. Buffalo were quite liable to do something similar, he informed us, and even the elephants might blunder into the guy ropes. 'Be prepared' he warned over supper that first night.

Just what sort of avoiding action we were supposed to take he was quite unable to say. There is relatively little you can do when a three-ton hippo chooses to roll over on a tent in which you are fast asleep.

We had just finished our meal in the mess tent that first night when a blood-curdling roar closely followed by a terrifying and all too human scream forced us to our feet in alarm. The mess tent was large and three-sided, open at the end facing the camp fire. With one accord Jonathan, Hugh and myself turned, grabbing our chairs, to face the entrance. A great deal of shouting filled the air and sounds of something crashing around was coming from the other side of the camp where our hired African camp attendants had their quarters. Clearly, a lion was in the camp and spreading panic among the Africans. Now, something was running towards us. We waited with bated breath and heaved a communal sigh of relief when the shaking figure of David, the head African, came

OPPOSITE LEFT *The camp*

OPPOSITE RIGHT *The following dawn was staggering*

LEFT *Fish eagle by Jonathan Scott*

BELOW *Wildebeest zigzagging over the horizon*

bursting into the tent. Thankfully a lion was not hot on his heels. 'Simba, simba!' he yelled in Swahili ('Lion, lion'), as if we did not know already. At the first sight of the lion, David's companions had fled in terror up the nearest trees, where they remained long after the animal had vanished into the night. All in all, it looked as though it was going to be an exciting filming trip. The marsh lions were obviously in residence. We hoped the wildebeest would make an equally impressive entry. We were not to be disappointed.

The sight that met our eyes the following dawn was staggering. In Africa the dawn chorus holds an element of threat. From the direction of the marsh a lion roared a fearsome challenge to the rising sun. All around us golden mist rose slowly in the exquisite light of early morning. Yellow-throated (long-claw) larks sang in the grass. Skeins of Egyptian geese and wood ibis flew noisily towards the marsh. The marvellous sounds of an African fish eagle competed with the loud piping of frogs.

A beautiful pied kingfisher hovered effortlelssly over the spring before swooping to pluck a wriggling catfish from the crystal-clear water. Great white egrets, and black-winged stilts and strange-looking hammerkops, waded in the shallows around the marsh's edge, disturbing huge whirligig beetles which danced to and fro on the calm water surface.

All was new and fresh. African dawn has the feel of the very first day about it. As the sun climbed higher, illuminating the distant escarpment, it gave life to an unforgettable scene.

Beyond the marsh, blackening the plains for as far as the eye could see, were the wildebeest. Thousands upon thousands were massed in endless columns, their long dark lines zigzagging over the horizon like some gigantic super-organism the vast herds eddied to and fro – an enormous black swell on the ocean of grass.

For eons the wildebeest, have come and gone across the Mara plains following their ancient

Wildebeest

more of the enormous wilderness than any other animals except man and the vultures.

The wildebeest were coming to the marsh to drink and then heading for one of their traditional river crossing points. We decided to skirt around the edge of the marsh and follow the herds to the river. It was only 7 a.m. and the lovely golden early light still revealed the glistening, dew-dappled spider webs that carpeted the long grass.

Less than two kilometres from the camp we came across our marsh lion pride. There were four adult lionesses and ten cubs but no sign of any males. The cubs were playing among themselves, and with the lionesses. All were oblivious of the nearby wildebeest. This lack of interest was explained by the half-eaten carcass lying nearby. They had killed during the night, probably several times, and were no longer hungry. Their fat bellies testified to that.

Play is a dominant feature of lion activity and is of vital importance to the cubs. Mock combat, like our boxing and wrestling, helps strengthen muscles that one day will be used in deadly earnest. Such behaviour also develops stalking skills. Lions are the only truly social cat and play helps shape the bonds of friendship that will ensure hunting success later in life. But fun and games only occur when the pride is free from pressure. Hungry lions and cubs abandon play and weeks may elapse without any friendly interaction. But the pride had fed during the night. For the moment the wary wildebeest had nothing to fear.

Life within the pride is filled with affection and tenderness. The lionesses and cubs, in particular, stick closely together and interact frequently in a friendly manner.

As we watched, one of the pride males appeared from beneath a bush and ambled towards the shade of a nearby acacia thicket. Males appear to be only grudgingly tolerated by the females. Despite their magnificent manes and aristocratic appearance, the male lion is in truth little more than a parasite upon the lionesses of his pride. Although an able killer when the occasion demands, he seldom hunts, happy to leave most of the strenuous work to the females. The main duty of the males is to preserve the security of the pride's territory, preventing disruption from intruders by scaring them off. Lions' faces are a constantly varying map of their emotions. Fear, curiosity, anger and warmth

migration routes, motivated by an insatiable appetite for lush grazing. Wherever they go a constant chorus of strange grunts accompanies them. Wildebeest are the least elegant of the antelopes. These bearded misfits in a family otherwise renowned for beauty and gracefulness lope along on rickety legs with a curious, ungainly, top-heavy gait, head and shoulders slumped forwards. But this awkward canter can carry them more than 48 km (30 miles) in a day. Zebra also take part in the migration, often joining the wildebeest columns, but in much smaller numbers and preferring to stick to distinct family units. By contrast the wildebeest herds are an amorphous, loosely mixed collection of individuals constantly changing in composition.

Radio-tracking work has revealed that animals found in the same troop one day may be more than 40 km (25 miles) apart a week later. Although the movements of the long columns are roughly parallel and apparently synchronized with one another, they do not march across the plains in strict formation. There is variety in the amount of movement and direction of travel of the various 'armies' of beasts. Subpopulations of wildebeest probably exist in the Serengeti rather than one enormous migrating mass.

George Schaller, whose study of lion ecology is a classic of its kind, described the Serengeti as 'a boundless region with horizons so wide one can see clouds beneath the legs of an ostrich!' In the course of their travels, the wildebeest probably see

can ripple across their expressions from moment to moment.

This male's face seemed to mirror his role in life. Haughty and indifferent, he strolled after the pride with an air of arrogance, secure in his might. He had recently been involved in a battle. The side of his nose was ripped open.

There is a great deal of internecine violence in lion society. Males win their prides through fighting prowess and then have to defend them against all comers. Rivals constantly invade the pride territory and engage resident males in combat. Lionesses also squabble among themselves, especially on a kill, where they often slash and bite their way to the head of the feeding queue. All the lionesses had distinctive marks of one kind or

TOP 'Our marsh lion pride'

ABOVE *Lion male with the remains of a wildebeest*

179

another. Torn ears, cuts, facial and body scars were evidence of another, violent side to their lives. Like their faces, the individual personalities of lions vary considerably. This was to become obvious during the ensuing weeks as we got to know our pride more intimately.

Some were very playful, some highly strung and irritable. Others were sleepy characters who seemed incapable of working up enough enthusiasm for a hunt. This type did little more than keep a watchful eye on the hunting behaviour of their keener associates in the hope of being provided with a free meal. But one lioness we were to meet that morning would prove to be the star of our film. An irrepressible hunter, she seemed always to be on the alert. Whenever we encountered her awake, her steely gaze appeared to be focused on even the most distant possibility of dinner. If wildebeest, topi or hartebeest strayed within reasonable stalking distance she would be off, slinking silently towards them through the long bleached grass

We first came across her in the company of her two sisters. The pride had disappeared into the cool shade of an acacia glade where they would doze away the remainder of the day. We were making our way towards the river and were hardly out of roaring distance of the sleeping pride when we spotted three lionesses walking closely together in the direction of the marsh.

According to Jonathan they were former pride members, sisters who had been kicked out when their father lost control to the present pride owners. Forced into a life of exile they now led a nomadic existence over the surrounding plains. The arrival of the migration had brought them back to that old haunt.

'Loony', as we eventually nicknamed the dominant sister, would supply Hugh and I with hours of amusement and yards of excellent footage. She was by far the most energetic and adventurous cat in the Mara. We filmed three successful hunts from beginning to end and witnessed several more that were impossible to film because of poor light.

All of these hunts involved Loony. Several times we saw her abandon freshly killed prey and immediately begin stalking again without even taking a single bite. Here was a killer indeed. Multiple killing like this suggests that hunting behaviour is

not necessarily closely controlled by hunger. But her sisters benefited from the extra harvest. Lions suffocate large prey. They lack teeth long enough to penetrate the thick muscles and neck vertebrae of a fully grown wildebeest.

After days fruitlessly staking out the seemingly constantly sleeping pride, or driving around desperately hoping to discover some exciting action, we realized the most sensible strategy was to stick close to Loony. Whenever we did see any

OPPOSITE ABOVE *'Looney'*

OPPOSITE BELOW *Stalking*

RIGHT *Lioness chasing*

disturbance among the wildebeest in the vicinity of the marsh she was usually in the thick of it. One morning we located her some distance away on high ground above the marsh. She looked in the mood for hunting as usual and we trailed along a sensible distance behind her. Three kilometres (2 miles) and 90 minutes later she killed a wildebeest after a lengthy stalk and a magnificent chase. And we got it all on film. During her three-kilometre hike she hardly deviated from the task of finding something to hunt. She paused only to utter tentative roars aimed in the direction of the marsh as if to advertise her presence and check whether she could enter the territory of her old companions. The slumbering pride never bothered to answer. Their bellies were stuffed with food. Most of their killing was done at night and perhaps this was the reason that the exiled marsh sisters were able to come and go so freely during the daylight hours.

Not far from where we made our first acquaintance with the marsh sisters we found their father in the company of another male. Jonathan had named the latter Old Man. He was one of three lions who had controlled the marsh pride years before when Jonathan had first begun to record their family fortunes. His two former comrades in arms were both dead. Now battered and bruised, he was well

past his peak. Once invincible king of the marsh, he had lost one eye and his pride to the claws of his present companion. Yet, astonishingly, these once bitter enemies were now close friends.

In turn the father of the three marsh sisters was expelled from the pride, sparking off a chain of events that culminated in the banishment of the trio. For both these old males the glory days were over, but they still hung around the periphery of their previous domain.

Lion society shows a considerable fluidity in the comings and goings of its members. The heart of their social system is the pride with its closed circle of adult females and their cubs. Many of the lionesses will spend their entire lives in the pride in which they are born, living in a fixed territory which is handed down from one generation to the next. But some females are forced out, usually when a new male arrives on the scene. A new lion taking over a pride may also kill young cubs that are not his own.

The young males have to leave and either on their own or in a group lead a nomadic life style until provided with the opportunity of joining a new pride by ingratiating themselves with the resident males, or by force.

Females cooperate in the care of the young,

which are looked after in what could be described as a crèche. Cubs may suckle from any lactating female and filling up may require trips to three or four females in succession.

Head rubbing is the main form of greeting within the pride. Tiny cubs will head rub whatever part of an adult they can reach. Secretions from the facial skin may be passed between them providing a family smell which reinforces their sense of comradeship. Pride males are content to let the females lead from one area to another. Their relative lack of involvement in the communal hunts probably increases the chances of success. Their heavy manes, like enormous busbies, are not easily camouflaged. By hunting together lionesses greatly increase their chances of bringing down prey. Communal hunts are twice as successful as solitary ones. Once a kill has been made the male's superior size and strength enable him to push aside the lionesses and gorge himself first. Its every lion for itself when it comes to mealtimes. On large kills the whole pride will eat well. But if prey is scarce and the available food only the size of gazelles, the cubs suffer and may perish of starvation. Adults are capable of feats of gargantuan gluttony and can pack away over 27 kg (60 lb) of meat in a five-hour period. Schaller's studies reveal that on average lionesses consume 5 kg (11 lb) of food per day. But much of what they kill is inedible and it has been calculated that each lion needs 12 adult wildebeest per year just to stay alive. About 6,000,000 kg (13,000,000 lb) of prey disappears down the throats of lions each year in the Serengeti.

By noon on that first day we were getting more and more excited. A lot was happening in the Mara and already we were familiarizing ourselves with the sort of material we would need to develop the story of a single day. But the stifling heat of midday was accompanied by a lull in animal activity. It was as though in the course of a day's journey through this wide ocean of grass we and all the other vessels sailing across it had entered the doldrums – windless and still. The very centre of the day. We watched and waited.

Dragonflies lay motionless, the delicate mosaic of their wings outstretched over patches of brown, dry earth. Fragile flying machines grounded in the furnace of the Mara, temporarily cut off from the marsh spring and its ceaseless flow of water.

Cheetah dozed on the top of termite mounds, heavy eyelids flickering warily as if they could not trust themselves to sleep. Zebra stood around silently in groups, heads entangled like a striped puzzle glimpsed through the heat haze. Even larks, beaks wide open, panted among the grass stems.

The pride slumbered on in the shade, a bundle of legs and interwined tails. Some lay on their backs radiating body heat from their white stomachs.

OPPOSITE *Head rubbing is the main form of greeting within the pride*

RIGHT *Cheetah mother and young*

BELOW *Some lay on their backs radiating body heat from their white stomachs*

Only the wildebeest remained afloat, drifting in the hot, flat calm. All else was in a state of suspended animation. But not for long. Soon cooling breezes began to stir the grass, gently blowing life back into the dead noonday landscape.

Once again the long lines of wildebeest took shape and started to stream off towards the river, a collective will driving them ever onwards. Dust clouds began to envelop the galloping herds. In hundreds, soon in their thousands, the wildebeest were approaching the river, piling up alongside the high mud banks which towered over the muddy water.

Like the massed ranks of a monstrous medieval army they crowded together, tottering on the brink, hesitant to enter the swirling rapids. In the water hippos floated; huge snouts and the great mass of their heads just breaking the surface. The loud snorts they made as they submerged mingled with the weird groans and grunts of the assembled wildebeest. The noise was deafening. Huge crocodiles slithered along the water's edge or lingered expectantly in the shallows.

Hundreds of enormous Ruhell's and gryphon vultures stood around; some drooped themselves

ABOVE *Like the massed ranks of a monstrous medieval army they crowded together*

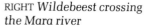

RIGHT *Wildebeest crossing the Mara river*

on the branches of the trees beside the riverbank, others squatted in a sinister fashion amid the nervous herds, their vast wings outspread. Like angels of death they watched over the multitudes — omens of impending catastrophe.

At first the wildebeest were very reluctant to cross but soon, driven on by the pressure from

behind, those at the front lost their caution and plunged into the fast-flowing water. The others followed.

In the treacherous swift currents the poorer swimmers struggled to get across. The advance guard had the easiest passage, but as more and more animals made it across the opposite bank became a steep slippery slope of mud on which the wildebeest began to flounder.

Soon the far bank was crowded with a heaving, jostling mass of beasts piled on top of one another in a frenzy of struggling bodies. It was an incredible and hopelessly sad sight.

In the middle of the packed, panic-stricken

throng, fights were breaking out and we could see many animals being trampled underfoot. Exhausted, some started to slip down the steep banks back into the water. It was obvious that many were drowning. Before long bodies began to float downstream, some on their sides, others upside-down with legs visible above the surface. Pathetic survivors attempted to turn back in midstream but only succeeded in creating further confusion as they collided with others desperately trying to cross. But there were still those who would profit from the macabre scene – crocodiles and vultures, for whom the bodies became floating dinner tables.

Even monitor lizards explored the possibility of a free feast. The monitor family includes the largest living lizards and we saw 120-150 cm (4-5 ft) long animals swim out to dead wildebeest trapped in the debris of fallen tree trunks and branches which littered the river bed. Monitors are excellent swimmers and use their flattened tails to propel themselves through the water. They normally feed on snails, insects and crabs, but carrion is a welcome supplement to their diet.

For countless years the wildebeest have acted out this tragedy on the Mara river. Why they risk such dangers remains a mystery. By far the

RIGHT *Before long bodies began to float down stream*

BELOW *It was an incredible and hopelessly sad sight*

majority survive the hazards of the river but in their wake the hippos are confronted with grim reminders of their passing in the form of dozens of bobbing carcasses. We witnessed hippos grab floating bodies in their huge jaws and throw them aside in fury. Perhaps an understandable reaction from a vegetarian whose living quarters are so badly polluted. Despite the dreadful carnage of these river crossings, the Serengeti wildebeest population has expanded from 2000 animals in the early 1960s to a present population of over two millions. Should the rains fail and the grasses wither they face starvation and a tragedy much more disastrous than the self-imposed sadness of these river crossings.

I turned away from the river overawed by what I had seen. Hugh had managed to film some details of the river crossing, but it was now late afternoon and time to be returning in the direction of the marsh and our camp. In one day we had observed much of what we wanted to capture on film. However, building meaningful film sequences requires careful consideration and patient effort. Three weeks of 12-hour days sweltering in the heat and staking out the pride were to follow. Our Land-rover would become a mobile hide out of which would slowly emerge 'Ambush at Masai Mara'.

As we approached the marsh we spotted the pride. In the coolness of the afternoon, the lions and cubs had moved from cover. In contrast to the lethargy of early morning they now showed keen interest in the wildebeest columns.

The dying sun was partly hidden behind gigantic thunderclouds which darkened the sky. But around the edges of the escarpment and in clear gaps between the heavy, black clouds, the last rays of the setting sun danced and dazzled.

At moments like this in Africa the light takes on an ethereal, almost mystical quality, illuminating individual lions in a golden halo. Brooding over the plains the stormclouds began to unleash cool rain showers, refreshing both wildebeest and lions. But the evening rain meant added danger for the wildebeest. Awake and alert, the lionesses were leaving their cubs to hunt once more.

For Hugh and I it had been a long, fruitful and exciting day. But in the course of the ancient evolutionary struggle that has united lion and wildebeest they, and all the creatures of the Masai Mara, had seen many such days.

6. STRANGER THAN FICTION

THE IMPOSSIBLE BIRD producer Richard Brock

Standing 2.75 m (9 ft) high and weighing over 135 kg (300 lb), the ostrich is one of the strangest birds in the world. Equipped with short fan-like wings on which it could never become airborne, the ostrich makes up for its flightlessness by being able to run at speeds of up to 65 km (40 mph) – faster than any other bird. Its stubby feet have only two toes, one being much larger than the other and capped by a broad claw which acts like the hoof of a horse when the ostrich accelerates into its loping stride. This claw is also a very effective weapon: a single well-aimed kick is enough to injure or kill a

LEFT *Ostrich breeding pair with two-week-old chicks*   ABOVE *Faster than any other bird*

OPPOSITE *Courtship display*

BELOW *It's the world's largest bird*

BOTTOM *Female ostrich feeding on grassland*

man, and even the lion is wary of this giant bird.

The ostrich, whose Latin name is *Struthio camelus,* is the only surviving species of a family of birds that 50 million years ago ranged as far north as Greece, Syria and Mongolia. Changes in climate and centuries of hunting by man have both worked against the ostrich. Scattered eggshells in the Middle East and North Africa show that the ostrich has disappeared from both these areas only recently, probably within the last 50 years. Today, it is found only in the tropical savanna region of Africa. This thinly wooded grassland and scrub provides ample food in the form of seeds, leaves and insect larvae, and has the advantage of being

open so that the ostrich can keep watch for its enemies, predators such as the lion and leopard. In East Africa particularly, the ostrich is still thriving and in Kenya it is possible to see the extraordinary breeding system that helps this bird to safeguard its chicks on the open plains.

The ostrich normally nests just once a year, usually between October and February. For up to five months before the egg-laying season male ostrichs claim and patrol territories which will help them to attract mates. A territory is the key to breeding success: any male who fails in the competition for land is certain to be ignored by females, and will have to wait a further year to breed. As the

time for nesting approaches, the males produce a booming song by inflating their throats and show off their plumage of jet-black feathers tipped with white on the wings and tail. Any rivals prospecting for land are quickly warded off; often a threat display with the wings outstretched is enough to deter another male, but if this fails, it is followed up with pecks and kicks until the intruder retreats. The months before nesting are a test of strength for the males and they feed enthusiastically, pecking rapidly at the vegetation until enough is gathered in the throat to be swallowed as a compact ball, known as a bolus, which can then be seen slowly travelling down to the stomach.

At the beginning of the breeding season the female ostriches wander through the territories of numerous suitors, inspecting not only the male birds but also the quality of the land that they have managed to claim. A territory containing many food plants is the most desirable, and the male that has been able to claim it is likely to be a healthy mate. As a female ostrich enters a male's territory, he immediately approaches the visitor to encourage her with a courtship display. As the plain brown female looks on, the male, whose bright red throat and legs contrast strongly with his black plumage, kneels to perform the 'kantle' display. This is a bizarre spectacle: crouching on the ground with his wings spread, the cock bird swings his long neck from side to side so that his head beats against his back, first on one side, then on the other. At the same time he makes a low booming call. When he stands up again his bright red penis leaves no doubt about his intentions. It is then up to the female to decide whether or not the male is suitable as a mate. If for some reason he does not sufficiently impress her, she will leave his territory and move on, perhaps to come back later if she has been unable to find a better suitor. But if the kantle display is successful, the hen's quivering wings indicate her interest, and the male follows her to a secluded spot in his territory where they mate. As in all birds, mating itself is a brief business: the female drops to the ground, and the male, making his subdued mating call, quickly mounts her. Within the space of a minute it is over, and immediately after this the pair may begin a bout of ceremonial feeding to reinforce the bond between them, both pecking rapidly at the ground with their heads held close together.

The search now begins for a nest site. The hen requires little more than a patch of ground scraped clear of vegetation, but the cock carefully selects a spot and marks it by scratching at the ground with his claws. Later, when the hen's eggs have developed, she returns to this nest site and enlarges it with her body before settling down to lay. Her eggs, weighing up to 1.4 kg (3 lb), are the largest of any bird yet, surprisingly, in relation to her body size they are among the smallest.

For the male, the annual round of courtship is not yet over. During the breeding season he may mate with perhaps half a dozen hens, although his first mate of the season – the major hen – has a permanent place as the dominant female. She is often the male's mate from the previous breeding season, and she takes an active interest in any hens that arrive in her mate's territory. Strangely, the major hen allows these subordinate females, or minor hens, to lay their eggs in her nest. She does not object to their intrusion and the clutch may grow from the dozen she has laid herself to as many as 50 or more – all fertilized by the same cock bird. The incubating ostriches have trouble covering so many eggs, and usually all but about 20 will either be pushed or roll out of the nest where they will fail

to hatch. The major hen's eggs are the least likely to suffer because, having been laid first, they will be in the centre of the clutch. It is possible that her eggs benefit from being shielded by those of the minor hens, as the eggs at the edge of the nest are those at the greatest risk from predators.

Ostrich pairs share the work of incubating the eggs, the female generally sitting during the day when the brown camouflage of her plumage is used to best advantage, and the male taking over at night. The female may stretch her neck out along the ground to improve her camouflage, possibly the origin of the inaccurate legend that ostriches bury their heads in sand.

The birds have good reason to keep the nest site hidden. During the six weeks of incubation the nest may be left unattended for considerable periods while the parents are feeding. Although the eggs are often partly covered by earth, predators are quick to find them. The Egyptian vulture is a specialist egg-eater which evolved the extraordinary ability to use stones to break the 13 mm (half-inch) thick shells. Picking up stones in its beak, the vulture drops them onto the eggs and,

although they rarely break first time, one successful throw is enough to provide a meal that will sustain the vulture for several days. Other predators like the cheetah are not so skilled at cracking the eggs: they try to bite at the shells or roll them on the ground until they break.

A few days before hatching the chicks can be heard calling inside the eggs. Unlike many other birds, the ostrich chicks have no sharp egg-tooth at the tip of the beak to help them crack the shell, and breaking out of an egg that can support the weight of a man standing on it is a laborious task. However, once hatched, like the chicks of most ground-dwelling birds, ostriches are well-developed. For them, just as for the young of the wildebeest, the zebra and other savanna animals, rapid development is essential to avoid predators. Within a few minutes of leaving the shell, ostrich chicks instinctively peck for food. After a short time the ineffectual pecking develops into proper feeding, but in the meantime they have enough egg yolk within them to last three or four days.

A pair of ostriches with several years' breeding experience behind them may hatch all 20 of their

OPPOSITE *The bright red neck and legs of the male ostrich during courtship*

RIGHT *The female ostrich stretches her neck out along the ground to improve her camouflage*

incubated eggs, but most pairs only hatch half that number. The newly hatched chicks are totally unlike their parents. Their downy plumage is fawn in colour with eight dark-brown stripes – an effective camouflage against dry grass and parched earth. They are about 30 cm (1 ft) high and initially grow at the rate of one foot a month.

As soon as hatching is over, the parents and their chicks abandon the nest to search for food. The chicks learn what to eat from their parents, choosing the greenest leaves, buds, flowers and seeds. As well as eating a large variety of plants they swallow small pebbles or stones which lodge in their crops to grind up their food, and in captivity they have been known to swallow metal objects as large as padlocks.

By one month the chicks are growing fast under the vigilant eye of their parents. The savanna is a dangerous place for the young birds, even though they can run at 55 km/h (35 mph). The sight of a bird of prey will rapidly bring the chicks together in a bunch while the adults flap their wings in alarm. By the end of the first year, only 15 per cent of the chicks will be left alive.

As the dry season comes to an end, families of young chicks escorted by their parents gather in the open. Breeding is linked with the arrival of the annual rains, so that the growing chicks are able to make use of the rapid flush of tender plant growth the rains bring. However, with the rains come new hazards. The filaments of ostrich feathers do not

OPPOSITE ABOVE *Newly hatched chicks*

OPPOSITE BELOW *Female ostrich and chicks in full attack display*

RIGHT *'Crèching'*

cling together like those of flying birds, and instead of repelling water, they soak it up. Ostriches also lack oil glands for waterproofing and a downpour can soak a chick, giving it a fatal chill. At the first sign of rain, the parent bird squats down and the chicks huddle underneath its body.

As ostrich families come together from their scattered nests, a remarkable event takes place, known as 'crèching'. When two broods approach each other, their adult escorts become tense and inquisitive. Suddenly, one male parent rushes away from his brood to inspect the other chicks. He does not walk up to them, but runs with his wings hanging low in an almost menacing manner. Having thoroughly investigated the brood, he returns to his own chicks. Then the process is reversed: the male from the other brood runs across in the same determined fashion. Without warning a fight breaks out as the adults struggle for supremacy. Weaving around the two broods, and nimbly leaping over the startled chicks, the adults clash in a contest of flailing claws and sweeping wings. Within a few minutes, the stronger male has established his superiority, and the subordinate adults leave, often followed by the winner's hen. Meanwhile, the chicks in the two broods have seen each other, and without hesitation they run together to merge into one flock in the sole charge of one dominant adult. The dispossessed parents have no more parental responsibilities, and are free to devote all their time to feeding.

The confused broods, or crèches, often consist of chicks of various sizes, because some have hatched a week or so before others. Eventually a crèche may be made up of perhaps half a dozen broods. This peculiar system of parental care ensures that fewer chicks are taken by predators: for plant-eating animals the best form of defence on the open plains is herding. This ensures that no predator can approach without being seen, and although some chicks may be killed, the majority will always escape. Thus, the ostrich crèche works in exactly the same way as the antelope and zebra herds with which they so often mingle.

The chicks grow at a prodigious rate. At nine months they are as big as their parents and can move just as fast. At the approach of the next breeding season, their adult guardian leaves them to set up a territory once again, and the young birds are left on their own. The chicks are fully grown within 18 months, but it is three or four years before they are fully mature. As they develop, they lose their juvenile plumage to take on the distinctive colours of the adults, and from then on they are treated as potential competitors by all other mature birds. In the space of a few years, the young males will be fighting for suitable mates. For many of the males, the struggle will be unsuccessful, but for those who do establish their superiority, the ostrich breeding system ensuresthat they are able to raise the maximum number of their own chicks.

L.G.

SCORPION producer Tony Edwards

'Scorpions?', colleagues gasped. 'You want to make a film on *scorpions?*'

It was the kind of reaction I was to meet again and again during my four-month acquaintance with these fearsome creatures. For scorpions seem to evoke in everyone the same mixture of fear, loathing and awe, even though most of us have never seen one.

I gulped my way through the entry in *Encyclopaedia Britannica*:

> **Scorpion.** A venomous arachnid of the order Scorpionida (about 800 species) ranging from half to 7 inches. [I later discovered this was wrong: there are actually 280 mm (11 in) scorpions!] . . . and having six pairs of appendages. The chelicerae, the small first pair, are used to tear apart prey. The pedipalps, the second pair, are large and have strong clawlike pincers, which are held horizontally in front and are used as feelers and for grasping prey. The last four pairs, each equipped with a pincer, are walking legs. The segmented abdomen tapers to form a 'tail' that has a stinger at the tip. Scorpions are nocturnal and feed mainly on insects and spiders. They grasp the prey with their large powerful pedipalps and tear it apart, sucking the tissue fluids; large prey is usually paralysed by the sting before it is eaten.

The entry under 'Astrology' was no more encouraging: Scorpio turned out to be a symbol of secrecy, sexual lust, darkness and death.

I saw my first live scorpion, however, not in any exotic tropical location, but in England, among the green pastures of Ongar, Essex. It is where May and Baker, the big agricultural and pharmaceutical

Fat-tailed scorpion

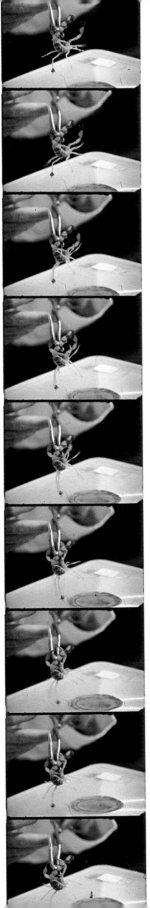

chemical company, has a research laboratory, and one of the firm's top insect men turned out to be Britain's 'Mr Scorpion'. His name is Bernard Betts, a fortyish-year-old bearded entomologist who has the rare skill of being able to look after and breed a whole range of tropical creepy-crawlies. His lab is alive with a terrifying collection of giant moths, deadly spiders and venomous ants – all extremely healthy, thank you very much. But Bernard's pride and joy are his scorpions, and when he talks about them his words spill out in a tumble of fascination, respect and – yes – even affection.

'People have got a completely irrational fear of scorpions. They think they are something that is deadly and that should be wiped out. But in fact most scorpions are completely harmless. Let me show you a few of my collection.' Suddenly, he thrust his hand into a box and pulled out a fearsome-looking black scorpion 230 mm (9 in) long. He was holding it by its tail . . . with his bare hands! 'This is the biggest scorpion of all – very large, very intimidating. It's the Imperial scorpion from Sierra Leone. It's got big claws for defence at the front which are quite capable of breaking a pencil. And yet it's very docile. The sting at the end of the tail is as mild as a bee sting, but to get it to attack you you would really have to provoke it.' He lunged at it playfully with a finger, but the scorpion hardly reacted.

Next he brought out Palamneus, a slightly smaller scorpion from Thailand and not quite so friendly. It made a grab at Bernard's finger with a large brown claw, but curiously did not use its sting at all. 'He won't use his tail unless I really provoke him, and I'm not going to do that', said Bernard. Palamneus meekly settled back in his box. I began to relax. The deadly scorpion was obviously a myth.

Bernard's manner suddenly changed. 'The next scorpion I'm going to show you I wouldn't attempt to pick up with my fingers.' Gingerly he poked a pair of large steel tongs into another box. Moments later the tongs had in their grasp a kicking, bucking, struggling mass of claws and legs. It was light yellow, 76 mm (3 in) long, and evil. 'This is one of the most dangerous scorpions in the world.'

FAR LEFT *Bernard Betts with Palamneus*

LEFT *The only way to handle a fat-tail*

197

Bernard held it at arm's length. 'It's called a fat-tail, and it's very aggressive. It will sting at the slightest provocation and the sting is very dangerous. The venom is a poison called a neurotoxin, it affects your nervous system. It can give you convulsions and paralysis, and eventually it will stop you breathing. Quite a lot of people have died because of this particular scorpion.'

Apparently the fat-tail can kill a human in seven hours, a dog in seven minutes, and a mouse in just a few seconds. Gratefully I watched Bernard lock the yellow peril away.

As the blood came back to my cheeks, he unleashed another excited volley of facts and figures about scorpions, which convinced me there was a fascinating programme to be done and that he was the man to front it. He told me about the scorpions' ability to withstand extremes of temperature (they will survive frozen in ice and in the hottest deserts). They can survive atomic radiation: in the French nuclear tests in the Sahara, the animals to withstand the most radiation were the scorpions. They can go without water for three months and food for nearly a year. A hardy lot . . . and yet there is a gentle side to them, Bernard said, which showed in their famous courtship and mating behaviour. 'How easy would it be to film?' I asked. He was not encouraging. 'Scorpions are nocturnal creatures for a start, and they're very sensitive to vibration. Get anywhere near them and they'll disappear under a stone or down their burrow – but it's worth a try.' He grinned.

Within an hour the plan was worked out. We would build an artificial scorpion habitat in Ongar, acclimatize the scorpions to our movie lights, and hope they would not mind doing it in public, so to speak. May and Baker kindly lent us a room, and a ton of earth, sand and rocks was heaved into a large deep wooden box. Bernard reckoned we would need at least two dozen males and females to have a good chance of finding a pair that would mate. But where were our movie stars to come from? Bernard knew the ideal spot – Morocco.

Thus it was that we were to be found a few weeks later in one of the most barren and desolate spots on earth. A totally flat, arid landscape strewn with football-size boulders, and with temperatures in the hundreds. 'Perfect scorpion country', said Bernard, as he marched off among the boulders. Every now and again he stopped to turn a stone over with his boot. Thirty boulders later a shout: 'Here's one!' A tiny scorpion lay huddled under the rock, and before it could scuttle away, the steel tongs had grasped its tail, and delivered it safely into a transparent plastic box. 'Buthus occitanus, very common in North Africa, and moderately dangerous.' He closed the lid. 'Our first male.'

It was orange-coloured, about 50 mm (2 in) long, and actually looked quite appealing. The brilliant sunlight showed up some fascinating details on its body. Very clear were the fine hairs on its front claws, designed to detect the faintest breath of movement in the still air and the prospect of a meal ahead. We could also see under its abdomen another of the scorpion's prey detection devices – the pectines. These look like a pair of tiny flexible combs, which the scorpion trails over the ground as it walks. The pectines are supersensitive; they can detect the minutest vibration and possibly even the chemical traces of prey in the vicinity.

By the end of the afternoon we had collected only five, but Bernard was not disappointed. During the day scorpions are difficult to find because they hide from the desert heat either under rocks or down burrows. The best time to look for them is at night when they all come out to feed.

Recently nocturnal scorpion hunting has been revolutionized by the use of ultraviolet light, – it makes scorpions literally glow in the dark.

We had brought with us a powerful ultraviolet lamp and it was an astonishing sight to switch it on in the pitch black of the desert night and see a mass of luminous shapes scurrying over the ground. Fortunately scorpions cannot detect ultraviolet light so we were able to get very close and film them catching their prey. A tiger moth, attracted by our lamp, fluttered briefly above a scorpion, but was suddenly plucked out of the air by an almost invisible snatch of a claw. We later studied the film frame by frame and discovered that the scorpion had grabbed the moth in less than a 25th of a second.

We returned to England with about 30 scorpions and our first duty was to declare them to H.M. Customs.

'I'd rather you put it back in the box,' said the customs officer, as Bernard dangled from his fingers a particularly vigorous specimen. 'Don't worry', he replied, 'it's quite docile.' The scorpion flashed its claws menacingly. 'I see what you mean', said the customs man with an odd grin. He had gone quite white.

Back in Bernard's laboratory the scorpions were introduced to their new home: a 90 cm (3 ft) square box with all mod cons, nice light sandy soil to burrow in, lots of rocks to hide under, and an equable mediterranean temperature. Around the edge, however, there was a glass-panelled 30 cm (12 in) high fence (just in case the scorpions did not fully appreciate what we had done for them). But, as it turned out, we need not have worried. The scorpions loved it.

During the next three months they happily set up home, went out hunting prey, danced their famous mating rituals and gave birth – all in the full gaze of Alastair MacEwen's lights and

Female Buthus occitanus *encourages a male to dance*

cameras. Alone and unprotected Alastair spent weeks cooped up with the fearsome creatures, and came away with some outstanding footage of the intimate details of their behaviour, some of it never seen before, let alone filmed.

Meanwhile, 8000 km (5000 miles) away in the West Indies, camerman Martin Saunders was also discovering what it feels like to be only a lens-width away from a scorpion. We had gone to Trinidad, an island where up to 400 people a year are attacked by the native scorpion, *Tityus trinitatis*. It is a 50 mm (2 in) long, dark brown, spindly looking horror with a vicious sting. Fortunately very few people die from it, but it can produce nasty symptoms: vomiting, stomach cramps and temporary paralysis. In common with most scorpion venoms, modern medicine has no specific cures, so it is not surprising that Trinidadians rely on folk remedies. The cane cutters have a particularly bizarre one: if stung, the victim must capture the scorpion, cut off its tail, roast the carcass and then eat it. Another cure is to drink a potion of rum and herbs into which a live scorpion has been dropped; the patient needs a strong stomach and a strong head – the rum is 80 per cent pure alcohol!

Scorpions are not deliberately aggressive; they tend to attack only when surprised. One of the commonest causes of accidents is when a scorpion happens to wander into your house at night. If it stays until morning it will follow its instinct to look for somewhere dark to hide away for the day. Shoes are a favourite spot – with obvious unfortunate consequences for their owner. While in Trinidad we thought this might be worth dramatizing as a sort of Scorpion Safety Council commercial. The shooting script required a scorpion to enter an old plantation house, go into a young woman's bedroom and climb into her slippers.

LEFT *and* BELOW *Female scorpions carry their young on their backs*

OPPOSITE *Climbing into 'the slipper'*

Hollywood would no doubt have built a mechanical scorpion for the task, but on a BBC budget we had to settle for the real thing. Regrettably, scorpions have not learned to read scripts, are totally untrainable, and cannot be bribed by the normal food-reward techniques.

In case anyone else is dotty enough to try repeating the exercise, here are a few tips discovered during six nerve-racking hours filming inches away from a lethal *Tityus trinitatis*. First, place your movie star in the fridge; cool for at least ten minutes. Do not pick it up (as the experts do) by the tail; it will get very angry and will need to go back in the cooler. Instead, take a very large kitchen spoon and gently ladle the scorpion into the hollow of the bowl. You will observe it immediately heading straight for you along the handle. Do not panic. Simply release your grasp and hold the spoon on the wrong end. By now the scorpion will have reached the other end of the handle and be heading back towards your hand. Shift your grasp back to the handle . . . and so on. When you are ready to attempt filming the first shot ladle the scorpion gently onto the floor and watch it like a hawk. If it appears to want to move faster than you can follow it on your hands and knees, put it immediately back in the fridge. Repeat all the above until the scorpion begins to walk at a

gentlemanly pace. Do not let it out of your sight. As soon as the shot is finished, and the scorpion looks like disappearing off the set, immediately scoop it up into the spoon. Recommence the see-saw along the handle . . .

Those were just the preliminaries. There remained the problem of persuading the scorpion

on cue to enter the house, walk along the floorboards, go into the bedroom and climb into a slipper – but that is another story!

In a film that had its share of surprises the most astonishing episode occurred in the most astonishing place: the far reaches of the London Underground's Central Line at Ongar. Bernard Betts had recently investigated reports of a colony of strange creatures found in a nearby disused goods yard and he took us down to Ongar station to look for ourselves. As we walked along the platform, a familiar animal suddenly scuttled across our path. It was 25 mm (1 in) long, had a pair of claws and a sting on the end of the its tail. It was, of course, a scorpion. 'Oh my goodness!' said the stationmaster when Bernard showed it to him. 'I never knew we had scorpions here. I suggest we don't convey this to the passengers, because we may lose the few we've got.'

Bernard identified it as *Euscorpius flavicaudis* (scorpion with the yellow tail), a Mediterranean species. But what were scorpions doing in England? The goods yard was the clue: they had probably arrived by ship from the Continent, stowaways in a consignment of imports. And it was not the first time they had been spotted in England. The Natural History Museum has a specimen of the same species found a hundred years ago at a port in Kent. They were thought to have died out, but in the 1970s a thriving colony was rediscovered on the same site.

The dockworkers have known about them all along, of course, and have got used to seeing them clambering over the walls by the quayside. We duly went down to film them, but Bernard made us promise not to reveal their location. 'These scorpions are truly amazing. They're found here in large numbers – thousands of them – and it's incredible they should have adapted to our English climate.' He picked one up and let it wander over his hand. 'Like most scorpions they are completely harmless, but I don't want people finding out about them. Some local authority bureaucrat would get to hear of it and spray the area with insecticide. People have such an irrational fear of scorpions . . .' He cradled *Euscorpius flavicaudis* in his palm with a look of real affection.

Oddly, at the end of my own four-month close acquaintance with scorpions, I found myself knowing exactly how he felt.

VAMPIRE *producer Adrian Warren*

All over the world there are legends about blood-sucking vampires, legends that can be traced through history as far back as the ancient Egyptians and Romans, and are a result of man's preoccupation with death and the supernatural. But in Central and South America, many thousands of miles from Dracula's castle in Transylvania, there are bats that turn grisly legend into cold fact. In parts of the West Indies, country folk believe that they can protect themselves against a supernatural being that drinks blood by sprinkling grains of rice close to windows and doors. The legend goes that before the creature can attack it must pick up every grain – by that time it would be dawn and, as everybody knows, by cock-crow all vampires must return to the safety of their lair. This West Indian vampire is usually an old woman who, at night, turns into a bat. The very meaning of the word 'vamp' is a woman, an adventuress who exploits men. According to a New York psychiatrist who studies fears and phobias, the evil image shared universally by bats can be linked to

The common vampire

old women who try to dominate their men, the kind of woman, one supposes, often referred to as an 'old bat'. Strange, then, that Count Dracula was a chap.

The West Indian legends, in Trinidad at least, must occasionally be reinforced by the odd person who gets bitten by a true vampire, a small brown bat about 100 mm (4 in) long that feeds exclusively on blood, usually of animals rather than that of men. These little bats were not, as is popularly supposed, the model for Dracula: it was the other way round. When European explorers first arrived in South America, they were fascinated by the blood-feeding habits of these bats and christened them 'vampires' after those restless souls from Transylvania who rose from their graves on dark nights. In fact, as a result, all bats have suffered from the stigma of Dracula ever since. True they are angular, leathery and hang upside down, and most only come out at night; however, not only are the majority of bats harmless, but they help us in important ways by eating troublesome insects, and in some parts of the world by pollinating plants. When the blood-feeding habit in South America was first connected with bats, people blamed the wrong one, the giant spear-nosed bat *Vampyrum spectrum* — quite understandable, since it is the largest bat of South America with a wing span of nearly 90 cm (3 ft), but it is in fact a meat-eater, often feeding on smaller bats than itself, which it catches on the wing. It does not drink blood.

There are three different kinds of true vampire bats: by far the most common is *Desmodus* which feeds mainly on the blood of mammals; *Diaemus*, somewhat rarer and prefering birds as prey; and finally *Diphylla*, a rare bat about which hardly anything is known and which differs little from its cousins except that it has very hairy thighs. One thing they have in common with each other is that they are just as at home on the ground as in the air; most bats, if put on the ground, flop around awkwardly with their wings half spread and have difficulty in taking off. Vampires, however, have elongated forearms and are extremely muscular; with wings neatly tucked in, they can run on the ground with ease and at great speed. When taking off they spring high into the air before spreading their wings for flight. The ability to run on the ground is of great value to vampires, allowing them to locate their prey on the wing then land

some distance away from it, finally stalking it from the ground. It also enables the bat to escape effectively after feeding when it may be too heavy with blood to fly.

Given that these bats only operate on the darkest nights, live in dark caves and carry potentially dangerous diseases like rabies, a 'private life study' of the vampire was a somewhat difficult assignment for a film. Understandably, when the idea was put forward it was met with scepticism and uncertainty, although with sufficient enthusiasm to approve an experimental filming trip to try to obtain a sequence of vampires feeding. No consideration at that stage was given to the problem of showing such a grisly spectacle on television; we did not even know if we would be able to film anything at all. Choice of location was important; if we chose the wrong one we might sit for ten days and not see a single bat. Trinidad seemed to offer the best opportunities, not because there are necessarily more vampires there than anywhere else but following serious outbreaks of rabies there in the 1950s and '60s, a bat control group had been formed toto monitor vampire populations and attacks on domestic stock. Consequently all the major vampire roost are now known.

As a base in Trinidad we decided to use a research station in the Arima valley in the northern range of attractively forested mountains. The station had originally been established in the 1930s by Dr William Beebe for the New York Zoological Society and had good facilities for keeping bats for observational purposes: indeed, much of the important original research work on bat radar or echo-location had taken place there. The building had been deserted for many years but renovation had begun prior to its being reopened for use by field biologists. But it was still charmingly ramshackle with creaky doors, cobwebs, old brass beds and its own colony of fruit bats, which flew through the main hall at dusk and at dawn: a suitable setting for a film about vampires.

The next problem was to find a friendly farmer with animals under regular attack. This was not easy: like their supernatural cousins, vampire bats tend to home in on some individuals but leave others mysteriously unmolested. Farmers with great numbers of cattle and other livestock were of no interest to us since it would have been to difficult to isolate, for filming purposes, an individual under attack. We needed to find a farmer with a smallholding, but most are careful to seal off their precious animals at night to protect them. However, we eventually managed to find a donkey that was being attacked almost nightly in an open stall. For several nights nothing happened; it seemed that while our presence was of obvious benefit to the donkey, our hopes of filming the vampires were beginning to fade. During our nightly vigils we would switch on the lights from time to time to see if anything was happening and we could only presume that this was creating sufficient disturbance to keep the vampires away. What we had not considered was the moon. So far the nights had been clear and brightly moonlit; on these nights vampires either avoid foraging or attack other prey in the darkness of the forest. On the first overcast night at the donkey's stall, it all happened, and it took us completely by surprise. A dark shape had moved in towards one of the

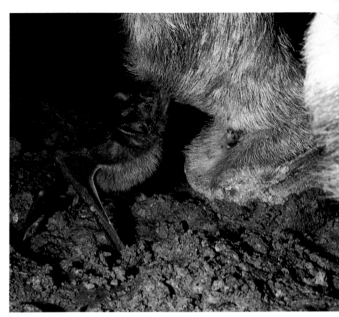

donkey's hoofs. When we switched on the light, it scuttled back into the darkness but to our delight came back again, at first slowly and cautiously, but then more purposefully. We were working with very low light levels and Martin Saunders, the cameraman, was only just managing an exposure with lens aperture wide open. As the bat moved in, Martin started the camera. Later that same night we filmed a second vampire attacking the donkey's ear. At four o'clock in the morning we returned to the creaky old house in the mountains bursting with satisfaction at having something on film. The following night we returned again to try for more; again the bats came and this time we were able to observe more closely their technique.

Before actually biting, the vampire often spends several minutes at its chosen site, sniffing and licking. Its saliva may contain a mild anaesthetic so that the bite might be relatively painless. The long fangs, the canine teeth so characteristic of the fictional Dracula, apparently enable the bat to shave some of the hairs at the chosen spot. Then it bites, using its razor-sharp incisor teeth to make a shallow scoop in the flesh and discarding the small piece of skin it has removed. The donkey appeared to be aware of the bite but not particularly worried. It shifted its hoof as if shrugging off a persistent insect and the bat hopped backwards out of the way before moving in again to lap the blood that

In flight

204

OPPOSITE *and* RIGHT *Hoves and ears are favourite sites for feeding*

BELOW *A donkey would have to be bitten a great many times to suffer as a result of blood loss*

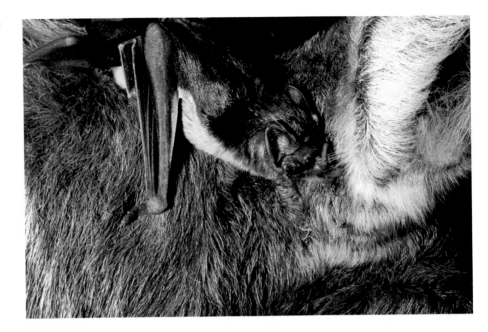

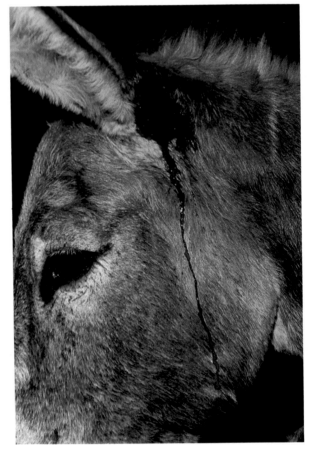

was gently flowing from the wound. Our lights, which were now on continuously, enabled the donkey, on occasion, to catch sight of the vampires as they scuttled in; at that he would shift nervously but most of the time he stood quietly. Once the bat was feeding, Martin was able to creep slowly in with the camera until finally he was only inches away from the bat using close-up attachments on the lens. The feeding would continue for as long as 40 minutes, the blood kept flowing by an anti-coagulant in the bat's saliva. In fact for quite a while after the bat staggered off with its belly full, the blood continued to flow; for such a tiny wound the loss of blood was large and for a smaller animal might have serious consequences to its health, but an animal as large as a donkey would have to be bitten a great many times to suffer as a result of blood loss. More serious is the threat of rabies, the virus of which is carried and transmitted through the bat's saliva. The work of the Trinidad bat group is to keep the vampire population down to a manageable level, destroying troublesome roosts and sampling others continuously to test for rabies. The statistics suggest that today, only one in two hundred vampires is likely to be a reservoir for the disease. I am glad to say that the donkey is still alive and well today, but the thought certainly crossed our minds that perhaps we should be chasing the bats off the donkey rather than settling

down to film the spectacle in all its grisly detail. It is difficult actually to admire any animal that feeds on another's blood, but perhaps that is not very logical. After all, we have considerable regard for countless other cruel predators like lions.

Over the weeks that followed the filming of the feeding, we began to view these bats very differently and even respect them in a strange way. The bat control group showed us several roosts to choose from for filming the next sequence, the bat 'at home'. One was quite spectacular, a limestone cave hung with stalactites and with deep chimneys in the ceiling. It was in the dark recesses of these chimneys that the vampires lived, hanging upside down from the rock in dense clusters. Feeding on blood may seem a squalid habit but the vampire is, like most bats, a sociable animal and keeps itself scrupulously clean, a large proportion of its time being spent in grooming, using the sharp claws of the feet for combing the fine fur. The fur is so fine that the Inca emperor Atahualpa in the sixteenth

A vampire cave attractively hung with stalactites

century is said to have had a cloak made solely from the skins of vampires. Even when excreting, the vampire meticulously pushes its body away from the cave wall in order not to soil itself. That was the unpleasant part for us standing underneath in a constant rain of bat excrement, which collected in pools on the cave floor. Unlike the guano of other bats, that of vampire is black and smells strongly of ammonia: two sure signs of a vampire roost. Not much, you would think, could live in the black pools that looked like sticky treacle, but we found several tiny creatures there.

Although the vampire's eyes are larger than those of many South American bats, they are of little use in dark caves so they use a kind of radar or echo-location to orientate themselves with their surroundings. They are also able to recognize other members of the colony individually and communicate with one another using a series of different sounds in the high-frequency range beyond human hearing. As well as sound communication, sniffing and grooming each other helps to form bonds between individuals. Courtship can be unceremonious, with the male often treating the female quite aggressively during mating. And afterwards, males play no further part in family life. A single young vampire is born after a gestation period of about two hundred days, nearly seven months. At birth it is only sparsely furred and blind; it uses hook-like milk teeth to cling tightly to the mother's nipple since, if it fell, it would certainly perish on the floor of the cave. When foraging, the mother usually leaves her infant behind in the cave where it is cared for by older youngsters and other females in a sort of nursery. There it is kept warm and safe and may even suckle from other nursing females if its own mother is away for a long time. The young vampire continues to suckle until it is nine months old, though during that time its mother will introduce it to a blood diet by first feeding it mouth to mouth, then by having it accompany her on sorties.

Despite our increasing admiration for these little animals, our particular roost was destined for the chop. Local farmers had been complaining of more and more attacks on their livestock. The bat control group had tried putting drops of strychnine around the site of the wounds, working on the theory that vampires often return to the same host and the same wound night after night. But this control method was not having a significant effect, so they tried a more lethal technique. The bats were caught in mist nets as they flew in to attack the farm animals; once one of the control team had removed a bat carefully from the net, another smeared petroleum jelly impregnated with poison on to its fur. The bat was then released to fly back to its roost. There, it begins to lick itself clean and so swallows the poison. Its companions, which always keep themselves so well groomed, come to help it clean up. So the poison spreads through the entire colony. The poison causes an internal haemorrhage and only a small amount is needed to cause death; it is estimated that for every bat caught, treated and released, as many as 20 others may die. The next day we visited the cave; of the 60 or so vampires we filmed only a few remained alive. The carcasses of the others littered the cave floor. We watched as the team collected the bodies

and placed them in plastic bags; if they did not do so some innocent scavenging animal might have come to feast on the dead bats and also die of the poison. It was a control method frightening in its efficiency, but although man is eliminating roosts around inhabited areas, small remote colonies in the forest continue to flourish. The fact is that before Europeans colonized South America and brought their cattle, horses and chickens, vampire roosts were probably far less numerous than they are today. Domestic stocks are easy prey for vampires and allowed them to flourish to an almost unlimited extent. In the natural state, vampires were probably few and far between, feeding on monkeys, peccaries, tapirs, anteaters and the larger game birds such as curassows. The vampire's way of life could be called a marvel of evolutionary specialization; it has few enemies, it does not tear its living prey to pieces like many of the carnivores we admire so much. Indeed it lives with great economic efficiency, only rarely killing its hosts; and were it not for the fact that – like our own much-loved dogs – it occasionally carries rabies, it could justly be called relatively harmless. But the name it has inherited means that in the eyes of most people this little, meticulously clean creature is doomed for ever to be loathed as a supernatural harbinger of death.

Chickens are also at risk with vampires around

ORINOCO HOG *producer Adrian Warren*

In the magical days of South American exploration, from the middle of the fifteenth century to the early nineteenth century, tales would drift back to Europe of perilous adventures, of the green festering hell of the jungle, and of strange animals: sloths the size of cattle; the anaconda, a snake big enough to eat a man; bats that feed on blood; a giant anteater with a long tubular snout, huge scythe-like claws and a tail like a parasol. Yet amongst this improbable menagerie, none was more odd than a giant guinea pig, the size of a sheep, with webbed feet and the bark of a dog: a beast made up from odd parts that could almost qualify as a mistake of creation? Charles Darwin's initial impression of this creature the capybara was disparaging: 'A curious animal of ludicrous aspect'; Alexander von Humboldt referred to it as the 'Orinoco hog'. In fact the capybara is not a pig at all: it is the world's largest rodent, an aquatic social animal living in packs and found throughout tropical South America, both in forest and on savanna.

In March 1800 Humboldt's explorations led him

The 'Orinoco hog'

ABOVE *Cayman basking in shrinking pool in the dry season* RIGHT *Llanos in the wet season*

down from the mountains of northern Venezuela on to the vast *llanos* of South America, the great flood plain of the Orinoco that stretches without a break from the Andes in the west to the Atlantic Ocean in the east, an area as big as Spain. But the German naturalist and explorer was not impressed. Beneath the burning sky, the earth was a mosaic of scorched cracks; the air was thick with the heat, nothing stirring but the dust devils that whirled across the desolate arid spaces. To Humboldt, the flatness of the *llanos* was depressing. As far as the eye could see there was 'not a hill more than a foot high', 'a vast and profound solitude that looked like an ocean covered with seaweed'. Huge herds of cattle and horses ranged over this landscape of devastating extremes, many of them to die of thirst during the long dry season, or to drown in the floods that follow the rains.

To find capybaras, essentially aquatic animals, in this harsh landscape today would seem to be something of a mystery, but even in the depths of the dry season a few muddy pools survive. As the water supplies dwindle, the small groups of capybaras merge to form herds of 50 animals or more. Much of the day is spent in sleepy torpor but occasionally in a sudden burst of activity one staggers over to munch a few blades of impoverished grass or shuffles laboriously to slump into the water to cool off. The immediate prospects for an exciting film about capybaras seemed slim, but these bizarre animals had more in store, and in any case the environment was exciting enough.

Under a clear blue sky in the dry season, the survival of the *llanos* community is tied to the dwindling water pools; as the water recedes through evaporation, the water and surrounding mud becomes overcrowded with cayman and turtles crawling over one another as they jostle for space. In very shallow pools, the backs of fish become exposed as they swim, now no longer able to submerge themselves, and their days are numbered. Storks and herons pace urgently near the water's edge, stabbing with their bills almost casually at the easy pickings.

In the distance, far across the plains, a line of trees shimmered in a mirage beside an illusory lagoon. We trundled along a dusty track in our jeep, exploring the *llanos* for ourselves, sharing observations and plotting the structure of the film. We stopped at a pool where 20 or so capybaras lay sprawled out on the mud. In response to our arrival one of them rose and made slowly for the water, its coat a mahogany russet and caked in dry mud that clung to the sparse lank hair. Still asleep were five large males, each weighing upward of 60 kg (130 lb), in the company of roughly twice that number of adult females. Among them were juveniles of various sizes, with lighter brown, sleeker coats looking rather more smartly dressed in comparison to the adults. In the very centre of the group a huddle of six babies snoozed together, three of them no more than 20 cm (8 in) long, the other three almost half as big again. Doubtless this sleepy heap of infant capybaras was the produce of two litters. Females give birth to an average of four young in each litter and, in the wild, they generally do so once each year. Births are scattered throughout the year, with a slight peak between April and June when the rains begin. Now it was February and the height of the dry season. The baby capybaras stirred and began a dance, each skipping

OPPOSITE *Young capybaras*
BELOW *Exploring the* llanos

into the air with a sideways flick of its rump. After each twitching hop the baby would glance at his companions as if to check that his antics received general approval. The playful babies churred quietly as they greeted one another and then turned their attentions to a female dozing nearby and began butting with their heads at her somnolent flanks. Try as they might the infants could not provoke any movement from the adult's corpulent form and so, their churring punctuated by aggrieved fluting whistles, the little mob turned its attention onto a different female. This one was more cooperative and stood up to allow all six babies to suckle. In addition to those on their flanks, capybaras have a pair of so-called axillary nipples, one in each armpit. So two of the nursing youngsters, one small and one larger, nestled into the female's armpits to drink. A few minutes later the female moved off towards the water, her activity prompting several other adults to follow. Among these was the female who had previously resisted the attentions of the thirsty babies. Seeing

her stand up, the six hurried across, and despite the female's attempts to walk, they soon drew her to a halt by their combined grip on her nipples.

These and other observations on females communally nursing were especially interesting to us, not only for the prospect of filming this activity but in learning something about capybara social behaviour. The communal care of youngsters may increase their chances for survival, and hence be of benefit to each of the mothers joining a nursing coalition; this offers a partial answer at least to the question of why capybaras are so sociable.

Male and female capybaras can be readily distinguished at a distance. All adult males share a feature that at first sight seems almost a deformity: they each have a large leathery blister on their snouts. In fact these glistening carbuncular disfigurements are scent glands. The females have them too, but theirs are only minimally developed in comparison with those of the males. The gland is called a morillo, from the Spanish word *morro*, describing a small, cube-shaped hillock. As male capybaras wander through their range they frequently pause beside a sprig of vegetation on which they firmly wipe their morrillo. As the gland is pressed against the shrub, a white viscous fluid is exuded. Next the male straddles the bush and urinates on the vegetation; and as he moves forward he scrapes the odour-daubed bush between his hind legs and dips down his posterior, probably depositing secretions from his anal gland. All in all these marking sites are liberally impregnated with odours that doubtless have considerable social significance. Exactly what messages these odours convey is uncertain and this is one question that is the subject of continuing study. Certainly different males would mark the same spots, perhaps trying to cover up another's scent with a new one and so asserting a dominant position on the social hierarchy. But, this is as yet uncertain.

From time to time adult males within a group would fight; no doubt for the capybaras this was violent social excitement, but to us they were meek exhibitions enacted in slow motion. For example,

211

one male may arise from his siesta and walk a couple of paces. A response may come from another male at the other side of the group who raises his head. The first one takes a pace forward and the second raises his head a further ten degrees. One more pace from the aggressor and the second capybara stands up too, another pace and the victim takes a stride in the opposite direction. Thereafter the two may shuffle a wavering course between the other capybaras, maintaining a roughly constant distance between each other, but triggering secondary shuffles by their progress as other males move out of the way. The frequent outcome was that the victorious animal succeeded in shepherding the other to the periphery of the group or even some distance away from it.

A similar sort of behaviour occurs between a male and a female during the preliminary stages of courtship. Some females may be promiscuous with several males within a group, but the dominant male takes precedence of choice. Once a female has been chosen by a male she is followed relentlessly; she may seem to play hard to get but the male is persistent in his courtship, staying close on her tail and frequently raising his head to savour her scent. His primary aim is to shepherd her into the water where all capybara mating takes place. Even in the water she may tease the male by giving him the slip, sinking suddenly below the surface and swimming underwater to emerge at an unexpected spot, leaving the male with head raised scanning the surface, waiting for her to become visible again. Then the following continues until she allows him to mount her. Like other rodents, mating takes place several times in quick succession, but the amorous liaisons tend to be short-lived; in contrast to the long foreplay it is all over very quickly.

One can only speculate on why mating takes place exclusively in the water. Certainly capybaras need access to water for it helps them to control their body temperature and it helps prevent their skin from cracking up through constant exposure to the hot sun. Other large, cumbersome creatures, like hippos for example, avoid strenuous activity on land where they would become quickly exhausted. In the searing heat of the *llanos* the strain

of mating on land might be totally shattering, leaving the animals less able to flee from a predator. So they do it in the water – the safest place since the capybara has no burrow to take his lady home to.

To film these scenes not only required a certain measure of luck but considerable patience on the part of the cameraman, Neil Rettig. On this open grassland it was not possible to stalk the capybara group without being seen, and since the groups tended to be relatively mobile, shifting their activities from one pool to another, a fixed hide was useless. So, together, we built a portable screen of split palm leaves on a frame of saplings which we carted using two diagonal supports. This plan also enabled us, over a period of several hours, gradually to shift the screen, equipment and ourselves nearer the group without arousing suspicion.

On one occasion, however, not only were we incredibly lucky, but we did not have to use the portable hide either. Jolting along in the jeep we spotted a group of capybaras bunched together on a muddy islet in a small lagoon. Through the binoculars, the capybaras seemed restless and we soon saw the reason: three rangy dogs waded through the water, splashing and probing among the beds of water hyacinth. The dogs turned their attention to the group of capybaras as we grabbed our equipment and scrambled along a dike to get a closer view. The next thing we saw was quite remarkable: the dogs, belly deep in the water, repeatedly lunged at the capybaras, but at each charge one or other of the adult capybaras lunged back towards the advancing dogs uttering a loud coughing alarm call. Each time the dogs retreated. Meanwhile, juvenile capybaras scrambled over each other's backs and squeezed through the adults to seek greater safety in the centre of the group, behind an encircling wall of protective bodies. As we brought the camera to focus on this scene, the dogs abandoned the attack and withdrew, followed by a hail of coughs from the capybaras. The filming opportunity seemed to have been lost, but then the dogs opted for a new tactic: they began to splash back and forth through the hyacinth beds, thoroughly quartering the area. Suddenly there was a splash and this time the camera was running. A juvenile capybara had not made the sanctuary of the group and had instead

Rangy (or feral dog) feeding on a capybara

crouched quietly and alone among the long hyacinth leaves. But the dogs had approached too closely, and it had broken cover. In a series of frantic splashing bounds it tried to escape but almost at once one of the dogs was upon the capybara and was soon dragging its dead victim towards dry land. The dogs ate communally and we were able to film them doing so. They are not wild dogs as such, rather domesticated dogs that have run wild, and are relatively common on the *llanos*. In fact these feral dogs are a major predator on the capybaras, whose only other natural enemies are jaguar, cayman and for youngsters, foxes and perhaps the opossum.

Although the capybaras may form tight groups to protect themselves, their main defence against predators is to leap into the water. They seldom stray very far from the water's edge and when danger threatens, their fusilade of coughing alarm calls is followed by loud splashes as the agitated animals dive to safety. For the first few days of life young capybaras are more reluctant to swim, however, and even older youngsters may tire when swimming across a large lagoon. It is then that cooperative behaviour again comes into action: adults will give youngsters rides on their backs. As a large female swims along as many as a dozen variously sized youngsters paddle frantically behind her in an effort to catch up and climb aboard. Smaller infants can travel long distances on an adult's back as it swims, but larger youngsters are inclined to have more trouble with their balance and roll off their mobile perch.

Man is another important predator of capybaras, at least in Venezuela where it is traditional to eat capybara meat during Easter week. The population of Venezuela is largely Catholic and traditional meats are forbidden at this time, but as the capybara is an aquatic mammal it is conveniently classified by some as fish and, therefore, legitimate Lenten fare. The annual slaughter takes place at a few selected locations where the capybara populations are high. It is all carefully controlled by the Venezuelan Ministry for the Environment who conduct a census to estimate the total population

LEFT *Adults will give young rides on their backs*

RIGHT *Scarlet ibis and egrets*

FAR RIGHT *Jabiru storks at their nest site*

214

Can they adjust to domestication?

before allowing the round-up to take place. The round-up takes about three weeks during which time about a third of the population is killed. Men on horseback drive the capybaras into groups, primarily selecting for males, and the death blow is by clubbing. But it is a time-consuming business rounding up wild animals; some biologists now suggest that capybaras could be efficiently farmed under controlled conditions in small enclosures, and experiments have been carried out to test the potential of this idea. The problem is one of adjusting a wild animal used to unlimited space to the kind of enclosure practical to farmers who may have little land and few resources. If this can be overcome then the capybara is a very promising commodity for domestication. Adaptation to captivity is good and females especially become tame. Their productivity compares very favourably with more traditional domesticated stocks. Free from the rigours of the extreme *llanos* environment, a single female in captivity will have two litters per year – a total of eight infants. Females can also be foster mother to other infants, further suiting its role as a farm animal.

It is too early to tell whether the idea will catch on, but it would be a bizarre destiny perhaps in keeping with the capybara's odd appearance. An extraordinary animal in the *llanos*, a dramatic landscape, as they lumber along the water's edge, the capybaras focus wildlife – lilytrotters, ibis and egret follow in their wake, pouncing on the tiny animals disturbed by the rodent's passage. Flycatchers even use them as a mobile perch. With its overwhelming flatness the *llanos* might be thought barren, but it is not. In one lagoon there were thousands of whistling duck, in another, brilliant, gaudy groups of scarlet ibis; in another, giant jabiru storks feeding and preparing a nest site. The list could go on but in our memory, one scene is more evocative of the *llanos* than any other: it is a pool glistening gold in the sunset, dotted with a mass of wading birds, amongst which, their heads strangely oblong in silhouette, wallow a group of capybaras – truly symbols of the *llanos*.

THE SERPENT'S SECRET _producer John Downer_

Upon thy belly thou shalt go and dust shall thou eat all the days of thy life.

Cursed in the book of Genesis for tempting Eve, the snake suffered a far worse fate than losing its legs – it was forced into the role of the most feared and hated of all animals. From that early story to the present day people have despised snakes and have justified that loathing by surrounding them with a unique mythology. Today, echoes of those beliefs are usually expressed by a few short words – 'slimy', 'scaly', 'creepy', 'revolting' – but they are just a legacy of those past prejudices.

At one time knowledge of animals was based on a mixture of random observation, hearsay and invention; against this background an animal that could move without legs, kill with one bite, and had also been pronounced evil by the Bible, gained an impressive but unfavourable mythology. It is only now that we are beginning to sort out fact from fiction and what is emerging is as fascinating as the legends.

In Britain we rarely see a snake, but in fact our islands are the home of the most successful snake in the world: the venomous adder. Known also as the viper, its range extends far greater than any other snake – southwards into southern Europe, eastwards as far as China and northwards right into the Arctic Circle. It shares our island with only two other snakes: the harmless grass snake and the rare and inoffensive smooth snake. Three snakes would hardly seem to justify legend but there are even ancient explanations why there are so few. The writer Pliny believed their numbers were kept so low by the mountain ash, a magical tree he thought would kill them. But the last ice age was in fact the real reason. As the ice receded only three species made it across to Britain before the landbridge to the Continent disappeared. St Patrick, for centuries credited with clearing Ireland of snakes, loses his glory to the same cold fact – Ireland detached itself from Britain before the reptiles had a chance to reach it.

Even without an ice age to contend with, Britain is not an easy place for a reptile; depending on the sun for their body temperature, life for them is a search for warmth. In winter the search is abandoned and they retreat underground to hibernate. Sitting out the winter in a state of torpor can be a risky business: the site must be free from frost and away from predators; suitable sites are scarce and many snakes may use the same disused mouse burrow or rotted tree stump. Stories of viper dens containing up to 40 animals are simply the discoveries of such popular sites.

The male adders are the first to emerge and may be found in early spring, basking in the sunshine. Like sunbathing holidaymakers they prefer a spot out of the wind and in direct sunlight, but they are not particular who they share it with and many snakes will crowd into the same prime spot. The

LEFT _Serpent tempting Eve_

OPPOSITE _Male adder (left), female (right)_

216

strange balls of snakes that people have reported are simply unusually large collections of these basking adders. All this sunbathing is vital for the adder and it is marvellously adept at it. It will change sites throughout the day following the movement of the sun, and is able to flatten its body to almost paper thinness to absorb every ray. As a 'cold-blooded' animal the adder is unable to maintain a high body temperature and in early spring the sun's warmth is needed for it to complete its sexual development and also for the first act after hibernation – the shedding of its skin. The adder will do this several times a year. The old skin is loosened by a secretion from the skin beneath and it is removed by passing through vegetation and pushing against rocks. It comes off in one piece, splitting first at the head then peeling backwards along the body like a sock being removed. The sloughed skin is a ghostly inverted replica of the snake that wore it. This habit enhanced the supernatural reputation of the snake, it was interpreted as proof of the immortality of serpents – the skin is shed but the snake lives on. Strangely one of our snakes does present an impression of immortality: a grass snake attacked by a predator or picked up in the hand will often roll on to its back, tongue lolling from its gaping mouth in a convincing impression of death. Left for a few minutes the snake will miraculously resurrect itself and crawl away: nothing supernatural but quite a useful defence trick against animals whose urge to kill is motivated by movement.

After sloughing, the males' colours are at their most striking: silvery grey or cream coloured with black zigzag marking, they contrast sharply with the red and brown hues and less distinct zigzag of the larger females. The females emerge from hibernation two weeks after the males and a few weeks later in mid-April there begins one of the strangest displays of any reptile – the so-called adder dance. Two snakes will suddenly rear up against each other and in a sinuous and unified movement they will begin to curl around each other, each striving to go higher and higher until with a violent movement they untwist and start again. For a long time it was assumed that this was some kind of courtship ritual, but in fact the participants in this serpentine ballet are all males. It is in fact a ritualized form of combat with the males competing for possession of a female. The aim of the dance appears to be to push the rival to the ground, but no sooner is one on top, than the other has raised his body that bit higher and has become the dominant one. Their upward gyrations tend to lead them to higher ground and they have even been observed climbing into bushes. The dance may

continue for over an hour, but eventually one of them will rapidly make an escape pursued briefly by the victor, normally the larger and fitter of the two. The victor then returns to the female to begin his courtship.

Sex and snakes were associated long before Freud made the connection, and this may well have contributed to their bad reputation. A mobile and living phallic symbol, the snake has often been portrayed as a seducer of women: the Devil, as well as having a serpent's tail, was often depicted with a highly versatile serpentine penis. This sexual connection is often exploited by striptease dancers who may use the snake as an unconventional partner in their performance. In nature the adder's courtship begins by the male rubbing his body against the female, at the same time licking her with rapid flicks of his tongue. The tongue is passed repeatedly along her body until she is ready to receive him. She signals her readiness by raising her tail and waving it slowly in a movement that is tempting to describe as a beckoning. The male now has an unusual choice before him – which of his two penises to use. The male has two penises; these are usually hidden within the snake, but when mating takes place they are turned inside out like the fingers of a glove and project out of the single opening. Only one is used at a time. They are spiny organs and once union has been achieved it appears that the male has problems freeing himself. Mating can last as long as two hours and the female snake often loses interest before the male, resulting in the unlucky animal being dragged along while still attached.

In the breeding season a snake killed by a blow on the spine may have his sex organs forced out, giving rise to accounts of snakes being found with two back legs.

All snakes produce eggs and in warm climates they are usually left to be incubated by the heat of the sun. With our unreliable weather such haphazard methods could be disastrous. The adder and smooth snake solve the problem by retaining their eggs in the body until the moment of birth. As well as protecting them from predators, it allows the female to keep them warm by seeking out the sun. Our only other snake, the grass snake, does lay eggs but uses a novel trick to provide extra warmth – it lays them in compost heaps and piles of rotting vegetation allowing the heat of decomposition to

LEFT *Adders entwined prior to courtship*

OPPOSITE *The belief that the female adder protects her young is a myth* speed up the incubation. Up to 50 eggs may be laid by each female. As suitable sites are scarce, many snakes may lay at the same spot and over 1000 eggs have been found at a single site. The hatching of such collections have given rise to periodic reports of plagues of snakes. The dependence on manure piles for incubation may have given rise to another ancient belief: that snakes could be spontaneously created by horsehair falling into liquid manure.

Even with the help of the heat of decay the grass snake is only able to breed as far north as the Scottish border, whereas the egg-retaining adder is able to breed in the far north of mainland Britain. Elsewhere in Europe it even survives above the Arctic Circle – further north than any other reptile.

The adder gives birth in late August or September to from 6 to 20 young. They are born in transparent sacs that break at, or shortly after, birth. The young are miniature editions of their parents complete with zigzag and fangs. Remarkably, it appears that they do not eat till the following spring, surviving the entire winter hibernation on food reserves gained while in the egg.

A persistent piece of folk-lore suggests that the adder protects her young and when danger threatens will swallow them. In fact the adder shows no parental care and the legend probably arose from snakes killed while giving birth; when opened up the living unborn young were found in what appeared to be the snake's stomach.

The fact that snakes can travel apparently miraculously without arms or legs has probably contributed to our mistrust of them. The Genesis story tells us the snake was forced on its belly through a curse, but in fact the absence of limbs gives the snake an important advantage, for it is able to move through dense vegetation and tunnel into the ground very easily. The snake has evolved from a four-limbed ancestor and it is believed that it lost its limbs as an adaptation to a life of burrowing. Many snakes have now returned to an above-ground existence but still have the advantage of a streamlined shape. Although no legs means that no snake is capable of speeds much faster than 5 km (3 miles) in an hour, for normal living they are able to move around quite satisfactorily. The adder moves in a typical serpentine glide by pushing the curves of its body against handy projections and thrusting itself forward, the

path of the head being followed exactly by the tail. As the snake moves, its tongue will periodically flick in and out. Although used as a symbol of deceit, the snake's forked tongue is in fact its most trusted sense organ. It is used to taste the air — useful for hunting down prey and detecting predators. The tongue relays scent particles back to the mouth where they are tasted by a special organ.

It was once believed that the snake could mesmerize his prey with a single glance, but those staring eyes are actually poor in definition although very sensitive to movement. The stare of the snake is that of an animal with no eyelids. The eye is protected instead by a layer of transparent skin which is shed periodically when the snake sloughs its skin. It is with these two senses of scent and sight that the snake does its hunting for it has no ears. The snake charmer's flute therefore goes unheard, the snake responding instead to the hypnotic movement of the snake charmer's body.

The snake swallows its food whole. The story that snakes can swallow food far broader than themselves is no myth. This seemingly impossible feat is achieved by a piece of brilliant animal design: both jaws are able to move independently and the surrounding skin is able to stretch to accommodate its victim. On one large meal the snake can go many weeks without feeding again and in captivity fasts of over a year are not uncommon. An active adder, however, usually feeds every one or two weeks. As the food has to be swallowed whole the kind of food a snake eats depends to some extent on how well it can control

ABOVE *The stare of the snake is that of an animal with no eyelids*

BELOW *The tongue relays scent particles back to the mouth*

it. The grass snake, for example, has no venom to overpower its prey, so usually feeds on harmless amphibians. The adder, however, uses venom to overpower its prey so is able to tackle small mammals, and these are the main part of its diet.

It is hardly surprising that the bite of the snake has been surrounded by myth: the prominent forked tongue was thought to deal the deadly blow; alternatively the sting was believed to lie in the tail. In reality the poison is injected into the snake's victim down two modified teeth—the fangs. These have a canal down their centre and, connected to the poison glands, they act like hypodermic needles. Usually they lie at rest horizontally in the upper jaw, but they can be quickly erected when striking at prey. If by chance the adder should bite itself no harm is done as the snake is immune to its own venom; so the old story that a captured snake would bite itself to death to save itself dishonour is therefore impossible. The actual strike is accomplished by a rapid thrust of the head. The bitten victim is immediately released and the venom is left to do its work. A small mouse takes one to two minutes to die and many actually struggle away from the snake in that time. Only after it has died will the snake look for it, tracking it down by scent, its flickering tongue guiding it to the spot. The effect of the venom is complex, attacking both the blood and nervous system; there is also evidence that it begins the process of digestion having been carried to every part of the body by the rapidly failing blood system.

Fortunately, its effect on man is considerably less lethal – less than ten people have died from adder bites this century and none in recent years. In Britain there is more chance of being struck by lightning or killed by an allergic reaction to a bee sting than dying through a snake bite, but such is people's fear of snakes that even a minor bite from one can cause severe shock.

In other parts of the world there are even recorded cases of people dying from the bite of a harmless snake purely through shock. In Britain the chances of getting bitten by an adder on a country ramble are very slight for it is a very unaggressive creature that prefers to flee rather than attack; the only time it is likely to strike is if it is cornered. A victim of snakebite should be rested and if possible the site of the bite immobilized: these precautions will delay the spread of the venom. Medical help should be sought and, most important of all, the victim should be assured that he will be all right. The old remedies of cutting the wound open, applying potassium permanganate and using a tourniquet will do more damage than the actual bite. In medieval times people believed that the only effective cures were potions made from parts of the snake's body – hardly surprising for at that time similar snake potions were believed to cure almost any ailment ranging from baldness to the pox.

The connection with medicine has had a long history and today the symbol of the medical profession is still the snake, based on the serpents of the Greek god of medicine Aesculapias. That it should have become such a powerful symbol of healing has probably developed out of the fear man has of snakes – an animal that can kill so effortlessly can surely heal as well.

But why are our reactions to snakes so strong? Is it simply the potent bite or is it the mysterious way the snake moves? Perhaps it is the sexual symbolism, or can it all be blamed on the Adam and Eve story? The true reason is probably a combination of these and just possibly one other factor – an innate fear. If we look at our closest relative, the chimpanzee, there is some evidence that this is the case but the experiments setting out to prove it have yielded mixed results. Chimpanzees brought up in the wild show a strong fear of snakes; those in captivity behave less predictably. The wild chimpanzee may have learned his response by watching how others react to a snake, but the reaction seems too strong for simple imitative learning. A possible explanation is that all chimpanzees have an innate tendency to fear snakes but it needs to be triggered by a confrontation early in life. If this is the case, in man it would certainly explain how our anti-snake society could so effectively create snake fears and phobias.

Whatever the reason, our view of snakes has changed little from those irrational fears of our primitive past, and many people will kill snakes on sight. In Britain more adders are killed by man than any other predator. It is to be hoped that, as we learn more about the habits of this secretive and inoffensive animal, such violent expressions of hate will be changed instead into a more rational feeling of respect.

ACKNOWLEDGMENTS

The publishers would like to thank the following:

Heather Angel, Farnham, Surrey: 34-35, 55, 157 (below right), 180 (below), 190 (centre left), 215 (above), 222 (below)

Ardea London Limited: 32 (both) John Mason, 33 (above) John Mason, 33 (below) Tony & Liz Bomford, 34 (left) Ken Roy, 35 (both) John Clegg, 37 Tony & Liz Bomford, 38-39 John Mason, 40 John Mason, 43 (left) Su Gooders, 51 Peter Steyn, 52 (below) Wardene Weisser, 70 (left) John Daniels, 70 (right) Ken Hoy, 76 Adrian Warren, 80 (above) Adrian Warren, 80 (below) M.D. England, 89 (above right) Clem Haagner, 89 (centre left) Su Gooders, 90 (below) L.H. Brown, 91 (above) A. Weaving, 92 K.W. Fink, 94 K.W. Fink, 95 C. Weaver, 96 (left) C. Weaver, 96 (right) Ian Beames, 97 (above) Ian Beames, 97 (below) P. Morris, 99 (below) Liz & Tony Bomford, 100 (above) A. Weaving, 100 (below) Richard Waller, 101 (above) Ian Beames, 101 (below) M.D. England, 103 (above) John Mason, 103 (below) Ian Beames, 106 Julie Bartlett, 107 (below) P. Morris, 132 Liz & Tony Bomford, 134 (left) David & Katie Urry, 134 (right) Peter Steyn, 135 (above) David & Katie Urry, 135 (centre left) Richard Vaughan, 135 (centre right) Richard Vaughan, 136-37 John Gooders, 154 (below left) John Daniels 177 (right) Clem Haagner, 179 (above) Clem Haagner, 189 Clem Haagner, 191 Clem Haagner, 192 Ian Beames, 193 J.L. Mason, 195 Ian Beames, 200 (below) Ian Beames, 202 Adrian Warren, 204 (above) Adrian Warren, 205 (both) Adrian Warren, 207 Adrian Warren, 209 (above) Adrian Warren, 211 Adrian Warren, 212 Adrian Warren, 214 (below) Adrian Warren, 215 (below right) Adrian Warren, 217 R.J.C. Blewitt, 222 (above) John Mason

Eric Ashby: 16 (below), 17, 18, (three), 19 (both)

BBC: 9, 11, 12, 13, 14 (both), 15, 23 (both), 45 (left), 46 (both), 47 (three), 48 (both), 49 (above), 197 (both), 200 (above), 201

Mike Beynon: 20, 21 (both), 22, 24, 25, 27 (below), 28 (both), 29 (both)

Stephen Bolwell: 49 (below), 168 (below left), 168-69, 169 (above), 169 (below)

Rodney Borland: 50, 54, 56-57

The British Library, London: 159

The Daily Mirror: 123 (both)

Neil Cleminson: 108, 109

Bruce Colman Limited: 146 Owen Newman, 148 Jane Burton, 149 John Markham, 150 Hans Reinhard, 151 (above) Kim Taylor, 153 (both) Jane Burton, 158 Hans Reinhard, 161 Hans Reinhard, 163 Hans Reinhard, 164 (below) Jane Burton, 165 Jane Burton

Susan Griggs Agency Limited: 78 Jonathan Blair 1981, 79 Jonathan Blair 1978, 84 Jonathan Blair 1978, 85 Jonathan Blair 1978

T.R. Halliday: 41 (below)

Mike Heywood: 26

David Hosking: 10, 41 (above), 68, 151 (below right), 152 (above), 156 (below left)

Eric Hosking: 73, 145 (above), 151 (below left), 178, 221

Martin Dohrn: 30, 36, 43 (right), 71 (above), 167 (below left), 167 (below right), 168 (above)

John Downer: 218 (right)

Mary Evans Picture Library, London: 216

The Kirkham Studios, Midhurst, West Sussex: 27 (above), 87 (both), 126 (below), 133, 176 (both)

London Scientific Fotos: frontispiece, 44, 45 (right), 166 (above), 167 (above), 170 (above left), 170 (above right), 171 (three), 196, 199

Mantis Wildlife Films, Australia: 59, 60 (both), 61 (both), 62, 63, 64 (both), 66 (both)

R.L. Matthews: 145 (below)

Nature Photographers Limited, Aldershot, Hampshire: 75, 160 Don Smith, 162 S.C. Bisserot, 164 (above) S.C. Bisserot, 218 (left) Owen Newman, 219 Owen Newman, 220 Owen Newman

Naturfoto, Denmark: 143

Owen Newman: 154 (above), 154 (below right), 155 (both), 156 (above), 156 (below right)

Robin Prytherch: 65

RAF Kinloss: 144

Professor H.B. Rycroft, National Botanic Gardens of South Africa 53 (both)

Michael Salisbury: Half-title, 110, 115 (above), 173, 179 (below), 181, 182, 183 (below)

Seaphot Limited: Planet Earth Pictures: 105 James Hudnall, 107 (above) James Hudnall, 111 Peter Scoones, 113 Jim Greenfield, 144 (above) G. Barker, 115 (below) Jim Greenfield, 116 (three) Jim Greenfield, 117 (below) Jim Greenfield, 118 Peter Scoones, 119 (below right) Jim Greenfield, 120 (above) and (below left) Jim Greenfield, 120 (below right) David Maitland, 121 (both) Jim Greenfield, 185 Keith Scholey, 187 Keith Scholey

Science Photo Library: 166 (below) Martin Dohrn, 170 (below left) Martin Dohrn, 170 (below right) Martin Dohrn

Jonathan Scott, Curtis Brown Ltd: 177 (left), 186

Keenan Smart: 176 (both), 180 (above), 184

Willy Suetens, Hever, Belgium: 138, 139, 140 (three), 141, 142

Maurice Tibbles: 71 (below), 124 (above), 125, 126, 127 (both), 128 (both), 129, 131

Scott Tibbles: 122

Adrian Warren: 77 (below), 81, 83 (below), 86, 88, 89 (above left), 90 (above), 91 (below), 93, 204, (below), 206, 209 (below), 210, 214 (above), 215 (below left)

Wildlife Picture Agency, Newton Abbot, S. Devon: 16 (above) Chris Peirce, 42 (both) Rodger Jackman, 74 Maurice Tibbles, 77 (above) Maurice Tibbles, 114 (below) Rodger Jackman, 119 (left) John Beach, 119 (above right) John Beach, 130 (both) Maurice Tibbles, 152 (below right) Rodger Jackman, 157 (above) Peter Smith, 183 (above) Ed Sadd, 188 Hugh Miles, 190 (above) Adrian Warren, 194 (both) Hugh Miles, 205 Adrian Warren

Rosemary Wise, Carterton, Oxford: 52 (above)

David Wright, Grimsby, S. Humberside: 8, 104